The Hook

The hook used to generate interest in force and motion is a thrill ride at an amusement park. News coverage of the opening day for two new roller coasters is used as the springboard for discussions that naturally include concepts related Newton's laws of motion.

Before playing the videotape, divide your class into the two- to three-member teams that will be working together on the task. Students will stay in these teams throughout the module, so assign them carefully. Examine the requirements of the task (see Teacher's Guide pages 2–3) for a guide to assigning students to teams.

Play the videotaped news coverage for your class. After viewing, each team should work together with another team to discuss what they have seen and to answer the five discussion questions below. Designate a recorder for each double team. The recorder will write the teams' best answer to each question on a large sheet of newsprint for use in a Blackboard Share session. Also designate a reporter. In Blackboard Share, the reporter's job is to report the teams' answers using the recorded list.

After all teams have reported their answers, the class should look for areas where they see agreement or disagreement. The sheets of newsprint can be saved and used at the end of the module as a way for students to compare what they knew about force and motion at the beginning of the module with what they know at the end.

Discussion Questions

1. Have you ever ridden a roller coaster? If so, describe the forces and feelings you experienced.

2. Is it harder to get a stopped roller coaster to move, or to keep it moving?

3. When you are riding a roller coaster—or in a fast automobile—that is rounding a sharp turn, you feel as if you are about to be thrown out of your seat. If the door were to fly open at that moment, and you were not wearing your seat belt, which direction would you go?

4. Which has more energy, a roller coaster almost stopped on top of the first hill, or one that is moving at its maximum speed at the bottom of the first hill?

THE STORY—PART 1
Fairs and Horses

Have you ever heard of the Beast, Rebel Yell, or Screechin' Eagle? How about the Corkscrew, Avalanche Run, or Magnum XL-200? They sound a little like television shows, or maybe books or movies, but they are structures that combine science, technology, and culture. They are thrill rides.

Hold on to the bar and keep reading.

Amusement parks started long ago. Archaeologists think the first recorded carnival was held in Egypt—a celebration of the receding of the Nile River's annual flood. Similarly, the Greeks and Romans set up markets at periodic times of celebration.

Fairs started in Europe during the twelfth century and flourished until the nineteenth century, when better transportation made goods available anywhere, anytime. The fairs often took place at the same time as religious festivals. Visitors to a fair came to sell and trade things.

To keep people at a fair, the organizers set up amusement centers where strolling musicians, puppeteers, jugglers, and dancers entertained the visitors. These outdoor events were held throughout Europe. The London Fair in England was one of the most famous. When fairs became hangouts for ruffians and criminals, however, they were stopped.

In the late seventeenth and early eighteenth centuries, pleasure gardens were built in Europe. Such places were more peaceful than the old fairs. Visitors walked through carefully planned and groomed landscapes. Some of the urban parks offered fireworks and balloon rides. Sales booths and tents dotted these early amusement parks.

Fairs started in the United States in 1811 when Elkanah Watson organized the Berkshire County Fair in Pittsfield, Massachusetts. U.S. fairs were more like exhibitions than markets, but they, too, had amusement centers.

Thrill rides had nonthrilling beginnings. The merry-go-round, or carousel, was the first. It can be traced back to twelfth-century war games held in Arabia and Turkey. Horsemen threw clay balls at each other. The balls contained stinking water. If the horsemen didn't catch the ball properly, they were fouled by the odor.

Kiddieland Carousel at Cedar Point in Sandusky, Ohio.

AN EVENT-BASED SCIENCE MODULE

THRILL RIDE!

STUDENT EDITION

Russell G. Wright

DALE SEYMOUR PUBLICATIONS®

The developers of Event-Based Science have been encouraged and supported at every step in the creative process by the superintendent and board of education of Montgomery County Public Schools (MCPS), Rockville, Maryland. The superintendent and board are committed to the systemic improvement of science instruction, grades preK–12. EBS is one of many projects undertaken to ensure the scientific literacy of all students.

The developers of *Thrill Ride!* pay special tribute to the editors, publisher, and reporters of *USA TODAY* and *NBC News*. Without their cooperation and support, the creation of this module would not have been possible.

Photographs: Pages 1, 4, 22, 28–32, 35–36, 46 Dan Feicht; page 16 Paramount Parks, Inc.; all Student Voices photographs Jeff Duncan

Managing Editor: Catherine Anderson
Project Editor: Laura Marshall Alavosus
Production/Manufacturing Director: Janet Yearian
Production/Manufacturing Manager: Karen Edmonds
Production/Manufacturing: Roxanne Knoll
Design Manager: Jeff Kelly
Text and Cover Design: Frank Loose
Cover Photograph: Terje Rakke, Image Bank

This material is based on work supported by the National Science Foundation under grant number ESI-9550498. Any opinions, findings, conclusions, or recommendations expressed in this publication are those of the Event-Based Science Project and do not necessarily reflect the views of the National Science Foundation.

ISBN 0-201-49737-9
Printed in the United States of America
7 8 9 10 11 12 13 14 - ML - 05 04 03

Dale Seymour Publications
Pearson Learning Group

1-800-321-3106
www.pearsonlearning.com

Contents

Preface

The Event-Based Science Model

Thrill Ride! is a module on force and motion that follows the Event-Based Science (EBS) instructional model. You will watch videotaped news coverage about two of the biggest, fastest, and scariest thrill rides in the country. You will also read authentic newspaper stories about thrill rides and amusement parks. Your discussions about thrill rides and amusement parks will show you and your teacher that you already know a lot about the physical-science concepts involved in these rides. Next, a real-world task asks you and your classmates to apply scientific knowledge and processes to the design of a thrill ride for a new amusement park. You will probably need more information before you start designing your ride. If you do, *Thrill Ride!* provides hands-on science activities and a variety of readings to give you some of the background you need. About halfway through the module, you will be ready to begin the task. Your teacher will turn you and your team loose to complete the task. You will spend the rest of the time in this module working on that task.

Scientific Literacy

Today, a literate citizen is expected to know more than how to read, write, and do simple arithmetic. Today, literacy includes knowing how to analyze problems, ask critical questions, and explain events. A literate citizen must also be able to apply scientific knowledge and processes to new situations. Event-Based Science allows you to practice these skills by placing the study of science in a meaningful context.

Knowledge cannot be transferred to your mind from your teacher's mind or from the pages of a textbook. Nor can knowledge occur in isolation from the other things you know about and have experienced in the real world. The Event-Based Science model is based on the idea that the best way to know something is to be actively engaged in doing it.

Therefore, the Event-Based Science model simulates real-life events and experiences to make your learning more authentic and memorable. First, the event is brought to life through television news coverage. Viewing the news allows you to be there "as it happened," and that is as close as you can get to actually experiencing the event. Second, by simulating the kinds of teamwork and problem solving that occur every day in our work places and communities, you will experience the roles that scientific knowledge and teamwork play in the lives of ordinary people. Thus *Thrill Ride!* is built around simulations of real-life events and experiences that dramatically affected people's lives.

In an Event-Based Science classroom, you become the workers, your product is a solution to a real problem, and your teacher is your coach, guide, and advisor. You will be assessed on how you use scientific processes and concepts to solve problems and on the quality of your work.

One of the primary goals of the EBS Project is to place the learning of science in a real-world context and to make scientific learning fun. You should not allow yourself to become too frustrated. There is more than one correct way of completing the task, and science activities can have many different solutions.

Student Resources

Thrill Ride! is unlike a regular textbook. An Event-Based Science module tells a story about a real event; it has real newspaper articles about the event and inserts that explain the scientific concepts involved in the event. It also contains science activities for you to conduct in your science class and interdisciplinary activities that you may do in English, math, or social studies classes. In addition, an Event-Based Science module gives you and your classmates a real-world task to do. The task is always done by teams of students. The task cannot be completed without you and everyone else on your team doing your parts. The team approach allows you to share your knowledge and strengths. It also helps you learn to work with a team in a real-world situation. Today, most professionals work in teams.

Interviews with people whose jobs are related to amusement parks are scattered throughout the Event-Based Science module, and middle school students tell of their experiences on thrill rides, too.

Since this module is unlike a regular textbook, you have much more flexibility in using it.

- You might read **The Story** of the event for enjoyment or to find clues that will help you tackle your task.

- You might read the **Discovery Files** when you need help understanding something in the story or when you need help with the task.

- You might read all the **On the Job** features because you are curious about what professionals do or to find ideas that will help you complete the task.

- You might read the **In the News** features because they catch your eye or as part of your search for information.

- You will probably read all the **Student Voices** features because they are interesting stories told by middle school students like yourself.

Thrill Ride! is also unlike regular textbooks in that the collection of resources found in it is not meant to be complete. You must find additional information from other sources too. Textbooks, encyclopedias, pamphlets, magazine and newspaper articles, videos, films, filmstrips, computer databases, and people in your community are all potential sources of useful information. If your school can link you to the World-Wide Web, the Event-Based Science Web page (http://www.mcps.k12.md.us/departments/ eventscience/) puts you in touch with timely data, images, and resources from around the world. It is vital to your preparation as a scientifically literate citizen of the twenty-first century that you get used to finding information on your own.

The shape of a new form of science education is beginning to emerge, and the Event-Based Science Project is leading the way. We hope you enjoy your experience with this module as much as we enjoyed developing it.

— Russell G. Wright, Ed.D.
Project Director and Principal Author

Fairs and Horses

Have you ever heard of the Beast, Rebel Yell, or Screechin' Eagle? How about the Corkscrew, Avalanche Run, or Magnum XL-200? They sound a little like television shows, or maybe books or movies, but they are structures that combine science, technology, and culture. They are thrill rides.

Hold on to the bar and keep reading.

Amusement parks started long ago. Archaeologists think the first recorded carnival was held in Egypt—a celebration of the receding of the Nile River's annual flood. Similarly, the Greeks and Romans set up markets at periodic times of celebration.

Fairs started in Europe during the twelfth century and flourished until the nineteenth century, when better transportation made goods available anywhere, anytime. The fairs often took place at the same time as religious festivals. Visitors to a fair came to sell and trade things.

To keep people at a fair, the organizers set up amusement centers where strolling musicians, puppeteers, jugglers, and dancers entertained the visitors. These outdoor events were held throughout Europe. The London Fair in England was one of the most famous. When fairs became hangouts for ruffians and criminals, however, they were stopped.

In the late seventeenth and early eighteenth centuries, pleasure gardens were built in Europe. Such places were more peaceful than the old fairs. Visitors walked through carefully planned and groomed landscapes. Some of the urban parks offered fireworks and balloon rides. Sales booths and tents dotted these early amusement parks.

Fairs started in the United States in 1811 when Elkanah Watson organized the Berkshire County Fair in Pittsfield, Massachusetts. U.S. fairs were more like exhibitions than markets, but they, too, had amusement centers.

Thrill rides had nonthrilling beginnings. The merry-go-round, or carousel, was the first. It can be traced back to twelfth-century war games held in Arabia and Turkey. Horsemen threw clay balls at each other. The balls contained stinking water. If the horsemen didn't catch the ball properly, they were fouled by the odor.

Kiddieland Carousel at Cedar Point in Sandusky, Ohio.

In France, the war game changed to a ring-spearing tournament. This horse-riding contest was called a *carousel.* To improve their hand-eye coordination for the carousel game, contestants practiced on wooden horses mounted on a circular platform. Servants, and later horses or mules, provided power to turn the apparatus.

By the eighteenth century, crude carousels appeared in parks as rides. You can find carousels today in most amusement parks.

Discussion Questions

1. Have you ever ridden a roller coaster? If so, describe the forces and feelings you experienced.

2. Is it harder to get a stopped roller coaster to move, or to keep it moving?

3. When you are riding a roller coaster—or in a fast automobile—that is rounding a sharp turn, you feel as if you are about to be thrown out of your seat. If the door were to fly open at that moment, and you were not wearing your seat belt, which direction would you go?

4. Which has more energy, a roller coaster almost stopped on top of the first hill, or one that is moving at its maximum speed at the bottom of the first hill?

IN THE NEWS

Theme parks let kids run a-muck

By Gene Sloan
USA TODAY

Parents, are you ready for this?

Starting Saturday, Paramount parks across the nation open new themed areas for kids designed to re-create the messy, hands-on children's TV shows on Nickelodeon.

Three-acre Nickelodeon Splat City opens first at Paramount's Great America in Santa Clara, Calif., then at Kings Dominion in Virginia (April 8) and Kings Island in Ohio (April 15). Like the cable network shows, they're aimed at 7- to 13-year-olds. "It's supposed to be an area where kids can run loose," says Paramount's Hugh Darley.

It looks that way. The attractions:

▶ **Mega Mess-a-Mania.** Get mega-messy at this live game show, with stunts, lots of audience participation

and elements from popular Nickelodeon shows such as *Double Dare* and *What Would You Do?* "We took all the mess that they've ever used on television and combined it," Darley says.

▶ **Gak Kitchen.** Help prepare new recipes from gooey Gak. Then taste such favorites as green slime slush and Gak *du jour.*

▶ **Green Slime Transfer Truck.** Watch the park "slimologist" conduct Green Slime experiments — and, yes, kids can participate. In case you haven't had enough, monitors on the truck broadcast clips from Nick shows.

Children can also climb through the Crystal Slime Mining Maze, get drenched under an erupting slime geyser and ride the Green Slime Mine-Car roller coaster.

"Bring a towel," Darley advises. And a sense of humor.

USA TODAY, 24 MARCH, 1995

Disney animal kingdom a mammoth undertaking

By Gene Sloan
USA TODAY

Central Florida just keeps getting wilder and wilder.

Disney World's announcement Tuesday that it will open a sprawling animal park by 1998 follows a long-popular safari attraction at Busch Gardens Tampa Bay and an earlier announcement by Universal Studios Florida that it plans a park that will include dinos.

But of the current and future attractions, Disney's $760-million Wild Animal Kingdom appears the most ambitious.

"It's a celebration of all animals that ever or never existed," Disney chairman Michael Eisner said Tuesday.

Hence, the 500-acre park will feature everything from giraffes and elephants to dinosaurs and unicorns — with the usual thrill rides interspersed.

Plans call for visitors to enter through a misty jungle bigger than a football field, then be taken into two mammoth themed areas:

▶ **Africa.** A car-like vehicle takes visitors for a safari, starting in a re-created African village and moving across a savanna for close encounters with herds of giraffes, zebras, hippos and elephants. Chase poachers and face obstacles like collapsing bridges.

▶ **Dinoland.** Disney designers' version of *Jurassic Park*. Ride through the Cretaceous Period past animatronic dinosaurs. Join a "mission" to save a dino. Visit a dino dig.

Another attraction: Conservation Station, where visitors

AP

DISNEY PLANS: Creative director Joe Rohde, left, CEO Michael Eisner and Roy Disney look over a model of planned animal kingdom.

learn how to protect animals and how the park works.

The park centerpiece — the size of EPCOT's familiar Spaceship Earth — will be the Tree of Life, carved with swirling animal forms and with a yet-unnamed ride inside.

Disney is also leaving room to add two more areas later — an Asian-themed safari with live animals and a mythological animal-themed land, says Judson Green, president of Walt Disney Attractions.

In the dog-eat-dog world of central Florida theme parks, Disney's new park is currently scheduled to open a year ahead of the previously announced *Jurassic Park* dino area at nearby Universal Studios Florida.

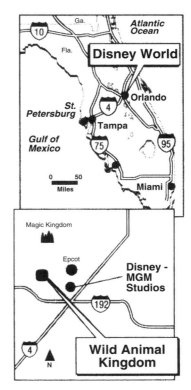

USA TODAY, 21 JUNE, 1995

Maximum Fear

Perhaps you have been to an amusement park. Some of the rides there are downright thrilling. They make your heart pound, your stomach rise into your throat, and your hands squeeze the closest person or object. Thrill rides give you a real rush!

One reason these amusement park rides are so much fun is the fear factor. For example, a roller coaster subjects you to forces you do not usually experience—unless you are a race car driver or astronaut. You are pitched, rolled, bounced, shaken, vibrated, lifted, dropped, and generally thrashed by the rides. If you are not used to that kind of treatment, you probably find it thrilling! You might also find it frightening at times.

Demon Drop at Cedar Point in Sandusky, Ohio.

The forces you experience on thrill rides were described almost three hundred years ago by a scientist and mathematician named Isaac Newton. After some keen observations, he wrote down his ideas about motion and gravitation. Today, we call his written observations *laws* because they are widely accepted as being scientifically accurate.

Newton's three laws of motion describe relationships between force and motion. The laws apply anywhere in the universe. They operate not only here on Earth, but also in interstellar space and on other planets. Newton's laws of motion can be measured on the Moon, in the space shuttle, in your family car, on your bike or in-line skates, and of course, on thrill rides. Newton's laws of motion are an important part of your everyday life. They affect almost everything you do!

Designing a Thrill Ride

A land developer is planning to build an amusement park on land outside your town. The developer wants visitors to the park to have fun, but she is also interested in the educational value of each ride. She wants her amusement park to demonstrate basic forces of nature—Newton's laws of motion.

The park needs to attract a variety of visitors and have safe rides. In order to get the best and safest designs for its rides, the developer has sent out a Request for Proposal (RFP) to companies that build amusement-park rides. You are a professional employee of one of these companies.

You will be working with one or two other students to design your thrill ride. As you design and build a working model of your ride, you may find it is too complicated to build the whole structure. If you obtain approval from the chairperson of the park's board of directors (your teacher), you may build only part of the ride. Remember, for your ride to be selected, it must illustrate Newton's laws of motion and it must be safe, thrilling, and fun!

At the very end of this module, you will be asked to vote for the best rides. These models will be displayed together in the actual layout of the amusement park.

Goround Amusements, Inc.
1928 Flume Street
Acceleration, NM 87543

Mary Goround, President

A Request for Proposals

We have received state and local approval for a new theme park on the outskirts of your town. We are looking for ambitious companies to design exciting new rides. The new park will offer rides that demonstrate some of the basic laws of physics. Special attention will be paid to Newton's laws of motion. The park will be called Sir Isaac's Inertialand.

Your proposal will be considered if you do all of the following:

1. Create a design for one ride that is fun and demonstrates Newton's laws of motion.

2. Give the ride a clever name that matches its target audience (young children, teenagers, adults, seniors, and so on).

3. Label the ride to show where it demonstrates each of the laws of motion (physics).

4. Create a sign to place at the entrance to the ride. The sign should explain, in simple language, the physics of your ride.

5. Create a brochure with a science activity that riders can perform while they are on the ride.

6. Give evidence that the ride is safe and will not subject riders to forces the human body cannot stand.

7. Write a newspaper advertisement for the ride. The ad should be designed for the target audience identified above.

Approximately one month from today, you and your colleagues will be invited to demonstrate a model of your ride to our board of directors. The board will make the final selection from all proposals submitted.

New Blizzard Beach is all wet and wild

By Gene Sloan
USA TODAY

LAKE BUENA VISTA, Fla. — Winter is on its way to Florida. Or at least sort of.

On April 1, the folks at Walt Disney World open Blizzard Beach, a whimsical $85 million water park themed as a wintery ski mountain that's, alas, *melting* (after all, this is Florida).

Among 25 water attractions:

▶ **Summit Plummet.** Hit 60 miles per hour on what's billed as the tallest (120 feet), fastest water slide in the nation.

▶ **Teamboat Springs.** Board six-person tubes for a 1,400-foot voyage — one of the nation's longest family whitewater raft rides.

▶ **Snow Stormers.** A slalom slide course through a forest of trees and "snow."

Another slide called Toboggan Racers lets folks compete side-by-side in eight lanes.

There's also a wave pool, a children's area called Tike's Peak and a tube ride through snow guns shooting water. To get back to the mountain's top, you can walk, or ride a ski lift — a water park first.

At 66 acres, Blizzard Beach is three times the size of the average water park. And the man-made mountain (six months and 50,000 truckloads of dirt in the making) is now the highest point in central Florida. It boasts room for 5,000 people a day.

The new park is expected to ease water park congestion at Disney World — the top destination in the country. Typhoon Lagoon, one of two other water parks at Disney, has been so popular that, because of crowding, it's had to sometimes close its doors 30 minutes after opening, officials say.

Blizzard Beach is just the latest in a flurry of construction at Disney. The company has already started clearing land for a fourth theme park, which should open later this decade.

Ticket prices at Blizzard Beach will be $22.50 for adults; $17 for children.

Walt Disney World

BLIZZARD BEACH: Disney's newest water park has 25 attractions, including an eight-lane slide that lets riders race.

USA TODAY, 24 MARCH, 1995

Indiana Jones' wild ride

Anything goes in Disneyland adventure

By Jefferson Graham
USA TODAY

ANAHEIM, Calif. — Indiana Jones is back. And Disneyland's got him.

The theme park today introduces the Indiana Jones Adventure, a four-minute ride through a world of skeletons, spiders, snakes and a huge boulder that almost falls on you — just like in *Raiders of the Lost Ark.*

Raiders executive producer George Lucas helped design the attraction, which took three years and nearly $100 million to complete. A new computerized technology called Enhanced Motion Vehicles offers slight variations on every ride. That's in keeping with the celluloid Indy.

"We said, 'What if we took Indiana Jones and action/adventure movies, and married them to a ride?" explains Disneyland producer Susan Bonds. "The movies are unpredictable — just when you think that something bigger, better or worse will happen, it does."

Disneyland's new ride, previewed on the Super Bowl halftime show, shouldn't be confused with the Disney/MGM Studio Tours Indiana Jones show in Orlando.

USA TODAY, 3 MARCH, 1995

Manager of an Amusement Park

WALT DAVIS
TOGO INTERNATIONAL, INC.
CINCINNATI, OHIO

I've always liked to build and create things. I studied civil engineering at Ohio State University. There I earned my bachelor's degree. Right after college, my first career was as an Air Force officer. My assignment gave me management experience, and the Air Force sponsored me for a master's degree in business administration. This added a mix of management skills to go with my engineering background. That combination turned out to be a magical one for me.

Fresh from 11 years of military work, I had an opportunity to work at Kings Island. Kings Island is a very popular amusement park in my hometown in Ohio. I began working there as director of capital expansion and maintenance. Capital expansion means obtaining money to help the park grow.

Moving from the Pentagon on one day to an amusement park the next was a big step. I suddenly found myself in the roller coaster business. On my first day of work, my boss gave me a big assignment. Build the world's largest and fastest wooden roller coaster! That turned out to be the world-famous roller coaster, The Beast. It opened in 1979.

Working at Kings Island meant handling some 20 major rides and another 50 medium to small rides. Each ride has a personality of its own. Most are custom designed. The bigger rides, especially roller coasters, are unique structures.

In this business, and perhaps in all jobs, you must have good communication skills and the ability to understand and get along with people. To me, those are the two most important skills in the workplace today.

It's good to understand the physics of how roller coasters work. How they are engineered might also be valuable knowledge. If you don't understand these things, surround yourself with people who do. In management, you can't do everything yourself. Find the people who have the skills you need.

In the winter, we can't run the roller coasters due to cold weather. There's too much friction. In addition, lubrication used on the wheels gets thicker in cold weather.

Kings Island has an extensive winter maintenance program. Winter is when all the coaster trains are taken apart. They are inspected and everything is double checked. Workers look for wear and cracks, and fix anything that is not right. The roller coaster structure itself is also inspected.

No two days are alike at an amusement park. On some summer days, Kings Island has more than 40,000 people in the park. There are days when rides have problems. They have to be closed and repaired quickly. Things can pop up fast.

A sudden change in weather can be a problem. Sometimes there are thousands of people in the park when the weather people call and warn us about possible electrical storms and 50-mile-per-hour winds. Where do you put 40,000 people in a hurry? We must have emergency plans. We have to alert everyone. Things can get fast and furious in that process.

I remember one very bizarre day. I had found a new type of pavement material for the walks in the park. It was an attractive, light-brown material. We wanted something different from black asphalt on our sidewalks.

I arranged to have the colored stone barged up to Ohio and sent in by train. A local contractor was hired to prepare test patches. That involved spreading tar, then covering it with the new material.

You won't believe what happened. Somebody jumped the gun. The contractor came in one morning two hours before the park opened and spread a layer of tar all over the place. On some spots the stone was put down. In other places, it was just sticky tar. I was attending other business that morning. I didn't know what was happening.

The park gates were opened. All of a sudden, my portable phone started ringing. There were problems all over the park. Visitors couldn't move! I went outside and couldn't believe my eyes. There was black gooey tar with visitors stuck in it. In some spots, just children's shoes were left in the tar. Smaller kids had run ahead of their parents and the tar caught them all. It was like flypaper. What a day!

If your task is to develop a park, there is a first priority. What will attract people to your park? Park visitors are always asking, "What's new?" You also don't want a ride that makes everybody sick. That's a bad investment. It can't be too rough or too scary. It has to be something people enjoy to ride. Certainly, the ride has to look attractive. Finally, will your ride attract enough people to pay for itself and make a profit for the owners?

Don't forget that different age groups have different ideas of what is fun. A teenager has a different notion of fun than a 55-year-old person. An amusement park must have something for everyone. Amusement park managers struggle to find the secret of what is fun for all ages.

In managing a park, you must be good at knowing what your visitors want. You talk with other park managers. You travel around and see which rides are more popular than others. You check out new ride openings. You identify what looks suitable for your park.

I'm now vice president at Togo International. We manufacture rides for parks. We're very proud of the feeling we have created with our new ride. We call it the *heartline roll*. Our coaster ride turns you upside down, but the maneuver doesn't

---

United States Patent [19]

Davis

US005433671A

[11] Patent Number: 5,433,671

[45] Date of Patent: Jul. 18, 1995

[54] **WATER AMUSEMENT RIDE**

[76] Inventor: Walter D. Davis, 5446 Hamilton Rd., Lebanon, Ohio 45036

[21] Appl. No.: **173,902**

[22] Filed: **Dec. 27, 1993**

[51] Int. Cl.⁶ A63G 21/18
[52] U.S. Cl. 472/117; 472/128; 104/70; 104/57
[58] Field of Search .... 472/117, 128, 129; 104/69, 70, 86, 56, 57, 73; 193/25 R, 27

[56] **References Cited**

U.S. PATENT DOCUMENTS

913,243	2/1909	Sonntag	104/57
3,635,185	1/1972	Kojima	104/56
4,805,896	2/1989	Moody	272/56.5
4,805,897	2/1989	Dubeta	472/117
4,836,521	6/1989	Barber	272/32
4,910,814	3/1990	Weiner	472/117

OTHER PUBLICATIONS

"The American Heritage Dictionary" p. 68.

Primary Examiner—Carl D. Friedman

Assistant Examiner—Kien T. Nguyen
Attorney, Agent, or Firm—Killworth, Gottman, Hagan & Schaeff

[57] **ABSTRACT**

A transport device is provided for transporting a water ride participant from a first elevation to a second, higher elevation. The device includes a spiral transport element extending generally between the first and second elevations. The spiral transport element has first and second end sections, an intermediate section and an inner surface extending along the intermediate and first and second end sections. The inner surface defines a spiral pathway between the first and second elevations. Further provided is a drive mechanism coupled to the spiral transport element for effecting rotation of the transport element such that the first end section of the transport element is capable of receiving a participant at the first elevation and the second end portion is capable of releasing the participant at the second elevation after the participant has traveled along the spiral pathway from the first elevation to the second, higher elevation.

11 Claims, 4 Drawing Sheets

create any centrifugal force. You hang upside down on restraints and feel negative *G* forces. That force makes you feel momentary weightlessness. We perfected this new sensation and it's extremely popular.

Science is everywhere in your daily life. You can't avoid it. Put on a coat in the winter. You're changing the thermal dynamics between your skin and the atmosphere. Use a knife. That's a wedge working for you. The tines of a fork also act as little wedges to help you penetrate food. Then friction takes over, holding the food onto the fork. That is science. Our life is inseparable from science—whether we switch on the lights or run the refrigerator. We are so immersed in science that science has almost become life itself.

Let me tell you about a new idea my son developed. He came up with an interesting concept when he was 16. The idea recently led to his first patent.

He and I were talking about water park rides. Water parks have had a big frustration—water always runs downhill. After their ride, people have to get out of the water. Then they

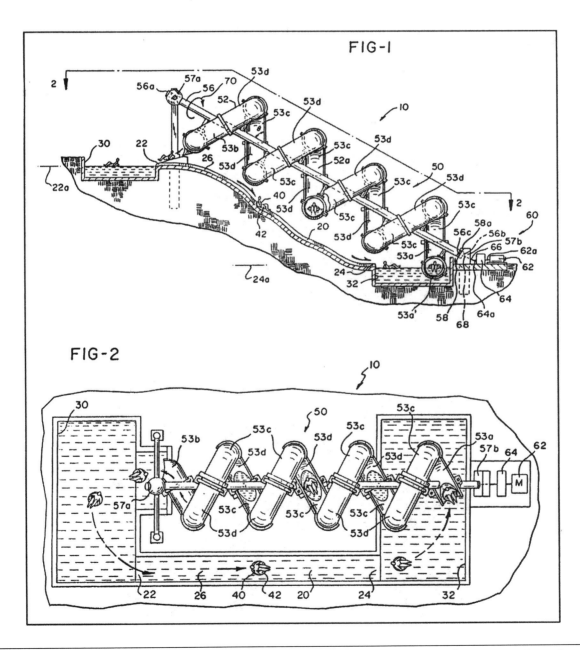

climb stairs to get back uphill. That interrupts their fun. Some water parks have tried pushing water uphill by brute force. Using big water pumps is very expensive and people don't like them.

As my son talked about the problem, he came up with a solution. Just use an Archimedes screw. He sketched it out and built a small working model using little Cheerios® for people in inner tubes.

The concept is to build a giant spiral out of fiberglass. You can then turn the spiral by electric motor. As it rotates, it moves people and water up the hill and puts them out at the top of the ride. The solution got him the patent.

A company in Canada has licensed his patent. The water parks see the idea as an inexpensive answer to their problem. And even better, it's fun to watch.

Walt Davis, Jr., with a model of his patented water return system.

Brake It, But Don't Break It!

Purpose
To use Newton's laws of motion to design a ramp that can transport fragile materials safely.

Background
You and your partners are industrial engineers working at a glassware company. At the warehouse, cartons of fragile glassware are stored in a loft above the ground floor. The president of the company has given you the job of designing a ramp for carts to carry boxes of glassware down to the main floor. The ramp must fit in a relatively small area and get the glass down quickly but without breakage.

The president decided it will take too long to wrap each glass with protective padding or to fasten the boxes down. Speed alone must provide the margin of safety.

Materials
For each group:
- Cart on wheels
- 2 or more wooden blocks smaller than the cart
- Long, flat board
- Meterstick
- 6 or more sheets of grid paper
- Assorted materials to slow the cart at the end of the ramp
- Thick books
- Masking tape

Procedure
1. Cut a piece of paper the size of the top of the cart. Trace the shape of one block on the center of the paper. The block must be smaller than the cart and free to move in any direction on it. Measure outward from the tracing and mark the paper every 0.5 centimeters in all directions. Draw an arrow down the middle of the figure. Tape the paper to the cart with the arrow pointing forward. During trials, the block should not move.

2. Place the block on the paper exactly where you traced it. Do not tape it down or do anything else to keep it from moving.
3. Create a table to record such data as track height, start and stop distance, number of blocks, and block movement. Each time you have a trial as explained in steps 5, 6, 8, and 10, record data in the table.
4. Prop up one end of the board on a stack of books or other support. You may notice there is a big bump at the bottom of the ramp. Design a way to lessen or get rid of this bump.

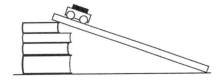

5. Start the cart at various positions on the ramp. Record each position as the distance from the starting position to the base of the ramp.
6. Release the cart. Measure and record the distance from where it starts to where it stops. Decide on a number of trials you think will show the relationship between starting positions and distances rolled. Keep track of any movement of the wood block.
7. On grid paper, plot the relationship between starting position and total distance. Do not connect the points.
8. Now explore the effect of varying the steepness of the ramp. This time, always start the cart from the same position on the ramp. Only vary the steepness. Record steepness as the height of the top end of the ramp. Record the total roll distance for each trial.
9. Use your data table to plot the points on grid paper. Do not connect these points.
10. Vary loads on the cart by adding blocks. These represent boxes of glassware, so you must not let the blocks move or fall off as they travel. Record your data on another sheet of grid paper.
11. Stop here and answer questions 1 through 4 in the Conclusion.

Now that you have finished the preliminary tests, it is time to prepare your design for the warehouse ramp. Your teacher will ask you to set up the ramp at a certain steepness. This time, you may use materials to stop the cart after it rolls down the ramp. The block should not move much after hitting the barrier. The group that does the following will win the contract with the glassware company:

- has the cart starting at the the highest position on the ramp
- stops the cart in the shortest distance
- moves the load the least
  Keep the block loose on the cart. The cart must run out (down) on a flat (not uphill) surface. After the competition is over, answer question 5 and the Extension.

Conclusion

1. Make a final copy of the graphs showing run distances related to ramp lengths and to ramp steepness (height). Choose a creative way to show the results of the load test. What effect does a longer ramp length have on run distance? What effect does a steeper ramp have on run distance?
2. What would happen to the blocks if the cart ran into a solid barrier? How does this observation show Newton's first law (the law of inertia)? (See the Discovery File

"Newton's Laws, a Moving Experience" on page 23.)
3. What kind of energy does the cart have before it starts to roll down the ramp? (See the Discovery File "Indestructible Energy" on page 17.) Describe the energy transformations as the cart rolls down the ramp and as the cart rolls along the flat surface. What is the cart's energy at the end of the run?
4. Describe the procedure you used to prepare for the competition. If your blocks

moved or fell off the cart, what was the cause?
5. How can the results of this experiment be applied to the design of amusement park rides?

Extension

In mountainous regions, road builders construct runaway truck ramps near the bottoms of long hills. These ramps allow a truck with failed brakes to leave the main road and coast uphill to a stop on a sand or gravel-filled path. Explain why the sand or gravel helps stop the truck.

STUDENT VOICES

I like the White Water Rapids because you get wet and there are lots of bumps and stuff. It's cool.

Thrill rides should have good features, like going very fast with lots of turns . . . things that will make you almost throw up.

CARLOS BACELIS
CULPEPER, VA

Gravity Rules

In the fifteenth and sixteenth centuries, Russian villagers popularized winter ice sliding in the outskirts of their towns. By the seventeenth century, ice sliders in St. Petersburg, Russia, had built a 70-foot-high framework of timber and logs. Thrill seekers sat in a block of ice, shaped to form a sled. They zoomed down the high incline on hard-packed snow. The rides were known as the Flying Mountains and the Russian Mountains.

Russian ice-slide managers made more elaborate and highly decorated ice sleds. They angled some of the slopes as steep as 50 degrees! They used sand to help slow the sleds. Slides quickly grew in length to several city blocks. Some ice-sled runs could accommodate up to 30 sleds.

The intoxication of height, speed, and sheer terror led to long waiting lines. Torches allowed the riders nighttime entertainment.

There were accidents, but despite the risk of being hurt, more and more daredevils were drawn to the ice slides. Survivors of these high-speed rides could boast of their courage and self-control.

It wasn't long before the sleds were fitted with small wheels so they would slide without ice, and roller coasters were born.

Most historians site the birthplace of the modern roller coaster as Paris, France, where a wheeled coaster ride called the Russian Mountains began entertaining Parisians in 1804. Accidents were common because carriages jumped off the track.

In 1817, Aerial Walks opened in Paris. Coaster cars rumbled down curved ramps on dual tracks, reaching speeds of 40 miles per hour. After each ride, attendants had to push passenger-filled cars up to the summit of the artificial hill to start the next run. Later, coaster cars were locked onto the track to prevent accidents, and a cable to pull the cars to the top of the hill replaced the pushing attendants.

Early roller coasters took various forms. One ride, called Niagara Falls, was much like shoot-the-chute rides. Passengers in a gondola were lifted into the air on one end of a giant seesaw. When released, the car sped down the slope into a river 60 feet below.

An engineer in Paris constructed the first loop-the-loop ride. Other loop-the-loops soon appeared. One ride, called the Centrifugal Railway, had a vertical somersaulting loop as part of its track. It opened in Paris in 1846. Released 30 feet above the ground, a car would run down a sloping 100-foot track. At the hill's end, the car would roar around a 13-foot-high circle of rail before stopping.

As a safety check, the Centrifugal Railway's first passengers were sandbags and monkeys. A workman then took a turn. At ride's end, he reported no trouble breathing during his looping ordeal.

Wooden roller coasters seem faster. The metal ones just kind of clank around. On wooden ones, you're going up and down hills, and you don't know what's coming next.

My favorite ride is the Grizzly. It's totally different from any other ride I have ever been on. When you get on the first hill, you aren't expecting anything, then you feel like you're going to come out of your seat. You have to hang on just to make yourself sit.

HOLLI DAVIS
CULPEPER, VA

Newton Rules

Sir Isaac Newton described many of the laws of physics that explain the relationships among forces, objects, and motion.

Newton's three laws of motion and law of universal gravitation help engineers design faster and more terrifying thrill rides that are also safer.

Isaac Newton was a pioneer in physics, the study of matter and energy. He was also an astronomer and a mathematician. He has been called the greatest scientific genius who ever lived.

Newton was born in England in 1642 to a farming family in Lincolnshire. He attended Trinity College of Cambridge University, where he studied the work of Nicolaus Copernicus, René Descartes, Galileo Galilei, Johannes Kepler, and other famous philosophers, scientists, and astronomers. He became a professor of mathematics at Trinity College in 1669.

By the time he was 24, Newton had already thought a great deal about forces between objects. One story says Newton got the idea about how gravity might work when he saw an apple fall from a tree. He wondered why the Moon did not fall from the sky the way the apple fell, and why, if it did not fall, it did not move away from Earth.

Newton did some calculating (much harder in those days without calculators). He estimated numbers, including how much pull the force of Earth's gravity might exert on the Moon. He knew he had to be more exact before telling people about his idea, but it was 20 years before he solved all the problems he encountered.

To see what drew Newton's curiosity, drop a pencil or other solid object. What do you think causes it to fall to Earth?

Newton reasoned that all objects attract one another. He called the force *gravity*. This single force not only pulls objects to the ground, Newton said, but also affects Earth's Moon and the planets. Newton showed in his theory how the entire universe is held together by gravitational force.

In 1687, Newton published his book *Philosophiae Naturalis Principia Mathematica*. (The book is usually referred to as the *Principia*.) In English, the title is *Mathematical Principles of Natural Philosophy;* like other professors of the time, he wrote his books in Latin.

Newton's laws were a triumph. His colleagues recognized his vision and reasoning power. Later, Newton laid the foundation for calculus, a branch of mathematics. His work also contributed greatly to the study of optics and color vision.

Using glass prisms, Newton discovered that a beam of sunlight (white light) passed through one prism split into beams of color that broadened as they traveled away from the prism. Newton then selected the blue light from the first prism and passed it through a second prism. He hoped this would help him understand how prisms produced the different colors. But, when he passed the blue light into the second prism, only blue light came out. This led Newton to propose the idea that white light is a mixture of different colors of light.

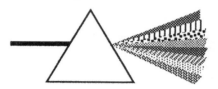

Although there are infinite colors that make up white light, Newton identified the red, orange, yellow, green, blue, indigo, and violet that we call the visible spectrum. He developed the laws of refraction and reflection of light and, in 1668, built the first reflecting telescope.

Queen Anne knighted him in 1705. He was the first scientist to be given the title. He was very proud of being Sir Isaac Newton. He is buried in Westminster Abbey, among the most famous people of England.

Isaac Newton created much of the basic foundation of today's physics. Single-handedly, he devised a framework for understanding the forces and motions of Earth and the heavens. Yet he honored Galileo, Descartes, and others when he said, "If I have seen farther than others, it is because I have stood on the shoulders of giants."

Roller Coaster Engineer and Project Manager

CYNTHIA EMERICK
TOGO INTERNATIONAL, INC.
CINCINNATI, OHIO

Here at Togo International, we manufacture roller coasters. I'm manager of the engineering department. My background is in material engineering.

My high school offered the usual blend of subjects, and I took them all. If you are interested in engineering, you need practical knowledge about just about everything. Take as many classes as possible. Every one is important, including shop. Knowing how tools are used is vital to my job.

In college, I studied engineering, but I received a lot of on-the-job training before coming to work here. My father is an industrial engineer and works with pneumatics and hydraulics. I worked with my father's business for many years.

Togo hired my father as a subcontractor to work on a prototype roller coaster. The company saw my skills in action as I helped my father on the project. Later, Togo hired me as a project manager.

No one engineer develops an entire roller coaster. It usually takes a team of more than a dozen engineers. The team comprises structural, electrical, mechanical, and biological engineers.

I often use my science knowledge outside work. Every winter, we have a lot of snow and ice. When my car gets stuck and the wheels are just spinning in the slush, I know there's no friction—nothing for the tires to grab. So I use a cardboard box. Putting cardboard underneath the tire creates friction. That's simple physics. It is surprising how much science we use daily.

Since I started working here, we've sold the first coaster that uses the heartline roll maneuver. Togo is the only manufacturer to build this type of ride segment. The heartline roll means just that. Your heart is in the middle, staying basically in one line. It's as if you were in a jet fighter doing a snap roll.

Some roller coaster rides have very big corkscrew rolls. Those have positive G forces. You feel several times the normal gravity on Earth. Centrifugal force is still pushing you into your seat.

Our very tight barrel roll makes you feel like you're falling out of your seat. The heartline roll produces a negative G force, so you feel like you're weightless, just hanging in the air.

The heartline roll was a tough engineering problem. Other manufacturers tried this same effect, but they didn't succeed. We had to engineer the track pipe in just the right way. We also had to design a car for the heartline roll segment. Of course, we had to keep the rider safe, and the rider's experience had to be fun. The combination of conditions was a challenge.

As a project manager, I am in front of the computer quite a bit. I review drawings. I decide what to buy and where to buy it. I also have to arrange for shipments to the construction area.

I often get calls from other engineers at various amusement parks. They ask for advice or have questions about one of our rides. On some days, I wear a business suit. Other times, I wear a hard hat, steel-toed boots, and jeans. That usually means I'm headed for a manufacturing plant. It is great fun seeing an entire project being completed.

As you work on building a ride, remember that the first priority is safety. Never forget that riders must be able to survive the forces you want to put them through. Decide what feelings you want your riders to experience. Do you want them to feel

Gravity's No-Show

By Dave Barry

weightless? What about positive *G* force? Answer those questions, then determine what physical forces are needed to produce the feelings you want. But never forget the safety of the rider.

We have a lot of ideas for the thrill rides of the future. Some ideas would be too expensive to develop now, but virtual reality rides are possible. Virtual reality rides make people feel things without actually moving; they tease the mind. Ideas about future thrill rides are often top secret.

Watching a train full of people get off a ride I designed is fun for me. They've been screaming with fear all the way through. Then they climb off the ride and say, "Let's do that again." That's when I love my job the most.

Torpedo water slide, Paramount's Kings Dominion.

SETTLE BACK, BECAUSE TODAY I'm going to tell you the dramatic true story of what happened when some Japanese researchers decided to recreate the historic discovery of the law of gravity:

As you recall, this discovery occurred in an English orchard in 1666, when, according to legend, Isaac Newton, the brilliant mathemitician, fell out of a tree and landed on an apple.

No, hold it, upon reviewing the videotape I see that in fact the apple fell out of the tree and landed on Newton. Had this occurred today, of course, Newton would have simply put on a foam neck brace and sued everybody within a radius of 125 miles. But those were primitive times, and Newton was forced to settle for discovering the law of gravity, which states: "A dropped object will fall with an acceleration of 32 feet per second per second, and if it is your wallet, it will make every effort to land in a public toilet."

Later on, Newton also invented calculus, which is defined as "the branch of mathematics that is so scary it causes everybody to stop studying mathematics." That's the whole POINT of calculus. At colleges and universities, on the first day of calculus class, the professors go to the board and write huge incomprehensible "equations" that they make up right on the spot, knowing that this will cause all the students to drop the course and never return to the mathematics building again. This frees the professors to spend the rest of the semester playing cards.

Yes, Newton made many contributions to science, but gravity was definitely his biggest. That's why a group of Japanese researchers decided, as an international goodwill project, to recreate the original discovery, using an apple tree that was descended from the original Newton tree.

I found out about this project thanks to an alert reader named (really) Harley Ferguson, who sent me a story about it from an English language Japanese newspaper called the *Daily Yomiuri.* The article states that researchers at the Construction Ministry's Public Works Research Institute in Arai, Japan, received a sapling descended from the original Newton tree. This sapling, according to the story, came from the U.S. Commerce Department's National Institute of Standards and Technology, or NIST, which is in charge of weights and measures. (So, if your pants don't fit the way they used to, this is the agency to complain to.)

I was curious as to why a U.S. government agency would be providing Newton saplings, so I called NIST and spoke with the official archivist, whose name (really) is Karma A. Beal. She sent me a bunch of information, which I will attempt to summarize here:

The original Newton tree—for simplicity's sake, let's call it "Bob"—died in 1814. But before Bob went to the Big Orchard in the Sky, cuttings were taken, and over the years these cuttings became trees, and cuttings were taken from those, and so now there are genetically identical offspring—let's call them "Boblets"—all over the world.

One Boblet lives at the NIST facility in Gaithersburg. It produced apples, but not many; the information Karma Beal sent me refers to the tree as (I am not making any of this up) "a very shy fruiter." The story gets a little murky at this point, but apparently the sapling sent to Japan for the historic recreation of Newton's discovery was grown from a seed from one of the NIST Boblet apples. This is significant, because if the sapling came from a seed, as opposed to a cutting, it is probably NOT a pure Bob descendant. As the NIST documentation states, "the original flower was almost certainly pollinated by some other tree."

But let's not be picky. The important thing is that the Japanese researchers had a sapling that was in some way connected to the original historic Bob. According to the *Daily Yomiuri,* their plan was to videotape the exact moment when the very first apple fell.

The sapling was planted, and eventually it produced a single apple. The researchers set up a video camera. All was in readiness as, day by day, the apple grew riper and riper, getting closer and closer to the big moment. And then, finally, it happened: A local resident, who knew nothing about any of this, wandered by, saw the apple, and ate it.

So the researchers never did get to videotape the apple falling in a historic manner, although the article states that "they did get scenes of the man munching on the apple." The man is quoted as saying: "It just tasted really bad."

But this does not mean the project was a waste of time. Often, in science, so-called "failures" produce the greatest discoveries. And this project resulted in a discovery whose value to humanity cannot be overemphasized. I refer, of course, to the fact that "Shy Fruiter and the Saplings" would be a great name for a rock band.

Indestructible Energy

Imagine this disappearing act. Energy that appears to vanish in one form must reappear in another form. It cannot actually disappear. That is against the law.

The transformation is described by the law of conservation of energy, the principle that energy cannot be created or destroyed. Energy can change forms, but does not cease to exist. All the energy at the end of any event is equal to the total energy before the event.

Consider a roller coaster speeding along on its tracks. As it moves along, it slows. It can never climb back up a hill as high as the first hill. In fact, the hills become lower and lower as the energy from the first hill is lost.

The wheels under the coaster car turn—a mechanical process. They become warm as they speed over the track. The warming is due to friction between the wheels and the track, and the axle and its bearings. Mechanical energy has been lost to friction. That lost energy reappears as heat, another form of energy.

Friction with the air also slows the coaster car. What happens to the energy lost to friction with the air?

On a roller coaster ride, energy also changes back and forth between potential energy and kinetic energy. Kinetic energy is the energy of motion while potential energy is stored energy.

A stretched rubber band, water at the top of a waterfall, and gasoline in a car are all examples of stored—or potential—energy.

From the top of a roller coaster's first hill, gravity pulls the cars down the track. Potential energy turns into kinetic energy—stored energy becomes the energy of motion. Rolling up the next hill, the cars start to slow. Kinetic energy is converting back to potential energy. Throughout the ride, energy is always conserved.

Putting on the brakes means turning up the friction. Flying down the tracks, coaster cars have loads of kinetic energy. Braking the cars demonstrates conservation of energy.

When brakes are applied to the car wheels, friction slows the wheels. As the brakes grab the wheels, the friction between wheel and brake converts the energy of rolling wheels into heat energy. If the brakes are applied long enough, the car comes to a stop. Kinetic energy of motion has transformed into kinetic energy of heat.

At the end of the run, the potential energy is gone—lost to friction and transformed into heat. Mechanical energy must be put back into the coaster cars again to haul the cars to the top of the first hill. Once at the top, the coaster has enough potential energy to keep the cars going to the end of the ride.

My favorite thrill ride is the Rebel Yell. When we were going down the first hill, I put my head down and it wouldn't come back up. I felt like the wind was holding my head down.

On a roller coaster, you can't tell the exact speed you're going, but you can tell when you're accelerating. The pressure builds up on you. You feel like you weigh more. I feel it mostly in my stomach.

PETRA LAWSON
CULPEPER, VA

Vice President of Maintenance and Construction

TONY RYLAND
PARAMOUNT'S KING'S
DOMINION
DOSWELL, VIRGINIA

I started here at the park in the mid-1970s when I was 16 years old. I had a summer job working in the ride operations department. Our job was to run the rides and keep the attractions safe for our guests.

Later I was offered a full-time job. I was in charge of the whole department of ride operations. About 10 years later, I was offered my current job. Now I'm in charge of running and building the rides as well as maintaining them. Through the years, my experiences have helped me take on the everyday challenges.

When I was in school, computers weren't available. But today, I would not be able to do my job without computer skills. We depend on computers in the office and out in the park. Even checking how our rides are operating, we use laptop computers. If your school offers classes in computers, be sure to sign up. Also take courses in math and science.

Don't forget, every job requires good communications skills too. Right now, I'm responsible for some 160 full-time people. About 85 of those people work directly for me. I can't do everything or be everywhere, so I communicate to others what tasks need to be done. I have to be on the same wavelength with my fellow workers. You must be able to communicate in order to give good directions.

At one time in my life, I was planning to be a school teacher. In a sense, this job is like teaching. We have a great team of people working here. And the work environment is very pleasurable. I wouldn't trade it for anything.

Here at the park, every day is interesting. There are always challenges. When we're open, we start early in the morning. Most days you'll find me working until 6:00 p.m. or later. A typical day starts by reviewing the previous day's problem reports. Ride problems have to be dealt with promptly.

Walking or riding my bike around the park is a daily task. I average around 15 to 20 miles during the day. As I walk, I check how the ride maintenance is going. It's good exercise, too. I usually get back to my office around the noon hour, just in time for staff meetings. Some days, there are budget reviews or discussions about future rides.

The science skills and knowledge I use at work I also apply at home. Right now, I'm building a new house. Reading the plans, changing the dimensions of a room, picking the right site—that's basic science. I play a lot of sports, too. There's a lot of science in keeping your body in condition.

It's great fun to see a new ride completed. One of my big challenges is to crank up a ride and make sure it works as we thought it would. When everything is finished and a ride is built, testing begins. At first, we use bags filled with sand or water. They simulate the weight of people. We run a ride for a day or so using those weights while we complete safety checks. We make sure the ride will stop and all restraints work properly. Often, we use an outside company to do final checks on the ride and the tracks.

Then it's my turn to be the test pilot! Many times, I'll take the first ride along with others in the maintenance and operations division. I'm not too nervous anymore. I've taken a lot of rides over the years.

We're excited about a new ride we're developing. It will use magnets to lift the cars rapidly to the top of the ride. The launching system is really different. Other roller coasters use other kinds of lift mechanisms, then let gravity do the work.

This new ride is inside a big, dark dome. There are lots of fast loops and four or five inversions. It's important to always come up with new ride ideas to keep ahead of the competition.

In designing your thrill ride, I have some thoughts for you. You need to target the type of guest you're trying to thrill. Is it an older person, a young adult, or a teenager? Your final park needs to be balanced.

Perhaps you want to design a wooden roller coaster. That means it will be difficult or almost impossible to do loops. On the other hand, a steel coaster allows you to do loops as well as various twists and turns. But above all, remember the safety factor. You must make sure your thrill ride is safe for the guests.

Don't forget the weather conditions. Is the ride indoors or outdoors? Consider the different kinds of weather in which a ride must operate.

Finally, be patient with each other. When you are brainstorming ideas, remember that nobody on your team has a bad idea. Work together. When we work on a project, it's not just one person who makes a ride work. Each person brings expertise. One person may offer talent in electrical engineering, another in mechanical hardware. Every person contributes. There's no weak link when you work as a team.

OLD-FASHIONED RIDE: The wood roller coaster Thunderbolt at Kennywood in West Mifflin, Pa., ranks second in a ride survey.

Try traditional Kennywood for turn-of-the century fun

By Gene Sloan
USA TODAY

If you long for the good ol' days at amusement parks, head to West Mifflin, Pa., outside Pittsburgh.

There, you'll find Kennywood (since 1898), voted most "traditional" by members of the National Amusement Park Historical Association. Despite upgrades, Kennywood has managed to save many of the old rides once common at turn-of-the-century parks, such as a rocking Noah's Ark ride, says Jim Futrell, the association's historian.

Runners-up in the survey were Cedar Point (Sandusky, Ohio) and Knoebel's (Elysburg, Pa.).

When it comes to roller coasters, however, the tables are turned and Cedar Point lands on top. The survey rated its Magnum XL200 and Raptor, a newcomer built last year at the 125-year-old park, the nation's No. 1 and No. 2 steel coasters. Kennywood's Steel Phantom is No. 3.

Among wood coasters, rated separately, the Texas Giant, at Six Flags Over Texas near Dallas, is tops. Kennywood's Thunderbolt ranks second.

USA TODAY, 11 APRIL, 1995

The Coney Island Connection

In the United States, coal mining and railroads sparked the thrill ride. Mining engineers needed reliable ways to move coal from point to point. Workers at coal mines laid tracks made of wooden rails topped with iron bars. Sets of coal-carrying cars were locked together at their hinges. Under gravity, the cars needed little more than a nudge to roll down a hill on the rails.

A notable development was the Switchback in Mauck Chunk (now called Jim Thorpe), Pennsylvania. It was the first gravity railroad in the United States. Originally built in 1827 to haul coal from eastern Pennsylvania mountains, the railway was abandoned in 1870. Townspeople converted it into a scenic tourist attraction. Passengers had panoramic views from the railway. Thousands of ticket-buying customers rode the Switchback. Today, the 16-mile railroad bed is a recreational trail.

The inventiveness of La Marcus Adna Thompson turned the Switchback railroad idea into a workable coaster ride. His ride opened June 13, 1884, at Coney Island, New York. The resort area was a popular summer weekend getaway for thousands of people.

Thompson's coaster was an instant hit with beachgoers. At five cents a ride, Thompson typically took in $600 a day. Sitting in cars holding 10 passengers each, riders experienced a scenic tour of the beach area at six miles per hour. They gently coasted on a 600-foot-long structure, which passed through a darkened tunnel (perfect for stealing romantic kisses).

Thompson's highly successful coaster at Coney Island led to the construction of more daring rides. In all, three amusement parks were built on Coney Island—Steeplechase, Luna, and Dreamland. Each park premiered thrill-ride attractions.

The Flip-Flap, for example, opened at Luna Park in 1900. The device put its riders through a 30-foot-high loop. The ride lasted 10 seconds. A passenger on this mechanical wonder was subjected to about 12 Gs of force. That's 12 times the gravity pull you feel standing on Earth's surface. Complaints of neck problems by some riding the Flip-Flap led to its early dismantling.

Yet another attraction at Coney Island was the Leap Frog. The thrill ride involved two electric cars full of passengers. Using the same set of tracks, the cars looked as if they would have a head-on collision. Just as an accident seemed certain, one car would race up and over the other on a set of curved rails.

Other coasters opened at Coney Island and Rockaway Beach, New York, as well as in Atlantic City, New Jersey, Ohio, and Massachusetts. Some coaster rides were deliberately engineered to include creaks and groans. Those sounds added more terror to the already fast-beating hearts of passengers.

Leap Frog Scenic Railway, circa 1920, at Cedar Point in Sandusky, Ohio.

Roller Derby

Purpose

To roll a ball down a track so it comes to a stop at a predetermined point, and to analyze energy transformations experienced on a slide.

Background

You are designing a slide attraction for a traveling carnival. For a slide to be safe, the rider must slow to a stop at the end of the ride.

You cannot take any chances, so you build a scale model of the slide. Carnival officials want to see your design work successfully before they give you their approval to build the full-scale attraction. They are also interested in the energy transformations that occur. The longer the slide, the more the carnival officials will like it!

Materials

For each group:
- Meterstick
- Strips of poster board, flexible white tag board, or other material to make a track
- Tape
- Support materials (books, blocks, pieces of wood)
- A few different balls (such as a marble, table-tennis ball, and golf ball)
- Small cup
- Blank paper

Procedure

1. Use poster board or other materials to build a troughlike track. Connect several sections together with tape. Be sure the tape does not interfere with a rolling ball.

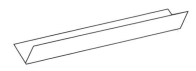

2. Set up the track so the starting point is at least 50 centimeters above the floor. Mark off in centimeters the section of track where you will start the ball rolling. Lay track elements end-to-end and carefully tape the sections together.
3. Position the track so the last segment of track runs uphill and the finish is a few centimeters above the floor. Place a cup under the end of the track.
4. Between the start and finish, the track can carry the ball downhill, uphill, or along a level path. The track can even curve, but somewhere along the way, it must touch the floor.
5. NOTE: Since this is a traveling carnival ride, the slide must come apart in several sections for transportation and storage. The longer your track, the more impressed the carnival officials will be.
6. Hold the ball somewhere on the upper part of the incline, and note the starting point. Let the ball roll down the track toward the finish line. If the ball is going too fast, it will shoot off the end of the track and miss the cup. If it is going too slowly, the ball will not make it to the end of the track.

7. Each time you roll the ball, record the starting position on the incline and note where the ball ends up. Keep adjusting until the ball falls into the cup at least twice in a row. Record the exact location where successful rolls began. Measure the height of this point above the floor. Also measure the height of the end of the track above the cup.
8. Try balls of different weights or sizes. Keep an organized data table.
9. Now it is time to demonstrate your attraction to the carnival officials. During your demonstration, you cannot touch or move either the ball or the slide. If it works, you are on your way! If it does not work, go back to the drawing board. If you have difficulty meeting this standard, others in the class will give suggestions as to how to fine-tune your model. Then you can make modifications and try again.

Conclusion

Since the carnival officials are looking at designs from several manufacturers, they want a written record of your design.

Include in the report a sketch of a side view of your slide. Mark the approximate starting location of the ball. Label where the ball's velocity is increasing, where it is decreasing, and where it stays about the same.

Whenever the velocity of an object changes, a force must be

acting on the object. On the sketch, draw arrows to show the direction of the force or forces acting on the ball in different sections along the track. Use different colors of arrows for the different forces (such as black for gravity, red for friction, and green for centripetal).

These questions will help you analyze changes in the ball's energy during the ride.

1. Before you let the ball go down the incline, what was the ball's kinetic energy? (Kinetic energy is the energy of motion.)
2. When the ball reached the finish line, it was hardly moving. At that point, what was the ball's kinetic energy (approximately)?
3. What is the height of the starting point (H_s) above the floor? What is the height of the finish line (H_f) above the floor?
4. What is the height of the starting point above the finish line? ($H_s - H_f$ = the vertical drop (D_v) of the slide.)
5. Where is the potential energy of the ball greater, the start or the finish?
6. If the kinetic energy of the ball is about the same at the start and finish, the loss of potential energy is due to friction. To find out the percentage of potential energy lost to friction, use this formula:

Percent Lost to Friction =

$$100 \times \frac{D_v}{H_s}$$

Set up of the experiment.

or Percent lost to friction is equal to 100 times the vertical drop divided by the height of the starting point.

7. If you covered the track surface with strips of sandpaper, how do you think the sandpaper would affect the ball? Would you have to change the starting position of the ball?
8. If you kept your slide as it is, but used a heavier ball, how would that affect the starting position of the ball?
9. Would it be possible to design a slide with a hill in the middle higher than the starting position of the ball? Explain your answer.
10. Small models of systems in physical science do not perfectly represent the real thing. In what ways is your model not similar to an actual slide at a carnival? Generate a list with your group, then write the ideas on a sheet of paper. Compare your ideas with those of other groups and expand your list.

Mean Streak roller coaster at Cedar Point in Sandusky, Ohio.

Newton's Laws, a Moving Experience

Isaac Newton was the first person to scientifically explain why objects move the way they do. He developed three laws of motion. His laws help explain the relationships among force, matter, and motion.

Newton's First Law

If you start a ball rolling, it will keep moving in that direction with the speed you gave it unless something stops it, slows it, or changes its path. Newton's first law says that until an object is acted upon by an external force, it remains in its current state. If the object is at rest, it will remain at rest. If it is in motion, it will remain in motion.

Newton's first law is called the *law of inertia.* Inertia is a property of matter. Inertia causes matter to resist changes in its motion. In his first law, Newton stated a new idea—that moving is as common a condition as resting.

The mass, or amount of matter, of an object is a measure of its inertia. All objects at rest resist motion. Objects with more mass resist motion better than those with less mass. For example, blow very lightly on a feather and it will move. Blow on a 50-pound steel ball as hard as you can, and it will not budge. The mass of the steel object resists being moved by the push of your breath. The steel ball has more inertia than the feather.

If you push hard enough on the 50-pound ball, it will move. Once moving, it will tend to keep moving, too. That is also because of its inertia. You will have to apply just as much force to stop the 50-pound ball as you did to get it moving in the first place.

Newton's Second Law

What happens to a ball when it is acted on by a force?

It will accelerate—increase in speed—in the same direction as the force. A steady force causes a steady increase in speed. Any sort of push or pull is a force. The greater the force, the faster the object accelerates. Also, the longer the force lasts, the faster the object will be moving when the force stops. That means that both small and large forces can accelerate an object to the same final speed. It just takes the small force a longer period of time.

Newton's second law, about force and motion, is usually written in this simple formula:

$$F = ma$$

It can be rewritten as:

$$a = F/m$$

In words, *force* equals *mass* times *acceleration.* Or, acceleration varies inversely with mass and directly with force. For example, pull back the plunger of a pinball machine as far as it will go. Newton's second law says a lighter ball will be traveling faster than a heavier ball launched with the same force. Can you think of a way to test this?

Newton's Third Law

If the pinball hits a wall, it bounces back. Why? Because when the ball hits the wall, it exerts a force on the wall. This causes the wall to exert an opposing force on the ball. The opposing force is always of equal strength but opposite in direction. The ball acts on the wall, and the wall acts on the ball. For every action, there is an equal and opposite reaction. This is Newton's third law of motion.

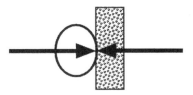

Newton's third law states that forces always occur in pairs acting in opposite directions and on different objects.

Law number three is all around us. Feel the recoil of a garden hose when you turn it on. Step from a boat and feel the boat move away as you step forward. Strike a nail with a hammer and see the hammer bounce as the nail hits back.

The next time you watch the space shuttle head skyward, remember you are watching Newton's third law in action. As the engines spew tons of matter from their nozzles, the matter pushes back. It is the reaction that pushes on the shuttle and forces it to rise.

Science Friction

Friction is nature's way of saying no. Newton's first law of motion states that an object in motion will remain in motion unless it is acted on by a force. Try this experiment. Start a ball rolling across a bare floor. Does it move forever, as Newton's first law suggests? If not, what is the force that slows the ball to a stop?

Friction is the force that resists motion. Friction is caused when a moving object makes physical contact with a surface—the ball contacts the floor.

Contact between a moving object and the medium through which it is moving also causes friction, such as a car moving through air and a boat moving through water. Friction is also caused by moving parts rubbing against other parts, such as an axle spinning inside a bearing. Friction pushes or pulls, slowing a moving object until it stops.

A roller coaster slows due to friction. When roller coaster wheels rumble over the track, friction between the wheels and the track opposes the coaster's motion.

Friction is also a destructive force. It causes wear and creates unwanted heat between the surfaces in contact. Coaster car builders sometimes use special material on wheels and rails to reduce friction.

Air also resists the roller coaster's movement. Friction caused by movement through the air is called *drag*.

When engineers build roller coaster rides, they take friction into account. Without friction, a roller coaster car released from the top of one hill could climb to the top of the next hill of the same height.

But friction causes energy loss. Friction transforms the mechanical energy of the turning wheels into heat energy. The heat energy is lost to the air.

At the top of the first hill, the coaster car has lots of potential energy. As gravity pulls the car downward, the potential energy becomes mechanical energy—a form of kinetic energy. Because the cars lose some of that mechanical energy to heat through friction, each hill on a coaster ride must be somewhat lower than the first hill.

The longer the ride, the greater the energy loss and the lower the hills have to be. The total energy of a ride does not increase or decrease, but the energy changes from one form to another form.

All this is not science fiction, it is science fact.

Mass and Weight

What is the difference between mass and weight?

Mass is the amount of matter an object contains. You have mass. Everything in the universe has mass. All objects have this property.

The mass of an object never changes by itself. The object can be on Earth, near Earth, or by a distant planet, and its mass remains constant.

The mass of an object is a measure of the object's inertia—resistance to change in motion. Inertia governs how an object responds when a force is applied to it.

Weight is the force gravity exerts on the mass of an object. An object has weight only in relation to gravity.

Different sizes of objects have different gravity. For example, the gravity on the Moon is one-sixth the gravity on Earth. If you weigh 100 pounds on Earth, you would weigh about 17 pounds on the Moon.

Weight even changes on Earth itself. A brick placed in a deep valley below the level of the sea will weigh very slightly, but measurably, more than the same brick placed on top of a mountain. In Death Valley, California, a brick is closer to Earth's center than it is on the top of Mount Whitney. The closer you are to the center, the greater the pull of gravity.

Ride Designer

**RON TOOMER
ARROW DYNAMICS
SALT LAKE CITY, UTAH**

I have a great job. Years ago, I would never have dreamed that my job would be designing roller coaster rides. It's been a ball. You just do one at a time. Then, pretty soon, you've got lots of them everywhere. Right now, we have nearly 90 coasters running. Our roller coasters carry close to 200 million riders every year.

I've always been a mechanic of one kind or another. In this job, you must understand how everything works.

In the early days, just out of high school, I was an automobile mechanic at a Ford dealership. In the early 1950s, I got drafted into the Army. Then I attended college and earned a mechanical engineering degree. It was in college that I picked up a lot that I had missed earlier. I took classes in physics, chemistry, and math.

Then I settled into an aerospace career. I worked on the rocket that launched America's first satellite, Explorer 1, in 1958. I also worked on the Minuteman rocket and the Apollo Moon landing program.

That was followed by a big dip in aerospace work. A friend of mine had been employed by Arrow Dynamics. He told me about this company and how they built amusement park rides. It sounded to me like the greatest thing. He introduced

me to the people who owned the company. They just happened to be looking for a mechanical engineer. I was the first one they had ever hired.

My first job here was to work on the company's first roller coaster ride. There were no promises of work after that one roller coaster was done. That was more than 30 years ago.

A typical day for me still involves a lot of original design work for coaster rides. I do legal work for the company as well. Writing proposals to sell our rides is another task.

When designing a coaster ride, the drawing board and a pencil are my main tools. I also use wire that I can bend into various shapes. The wire helps me imagine a new turn or upside down part of the ride. When drawing or working with wire, my imagination puts me on the ride.

Real coaster riding isn't for me anymore. But I've ridden enough of them over the years. I know what sensations are being produced. You get that seat-of-the-pants feeling after all those rides.

It is typically one year from the time a customer orders one of our coaster designs until the time the ride opens.

Engineers here at the company like to use computers to design rides. I think using computers to do your design work can sometimes slow you down. My advice is, don't forget the hands-on, practical side of mechanics.

Science has always been of interest to me. Understanding the weather, astronomy, and what makes the Earth work—these are special pursuits of mine. Understanding what things are all about is very important to me.

In designing your thrill ride, there are several considerations. First, take into account how much land you have to work with. Don't forget to adjust the curve radius of the track to the velocity of the coaster train. If you don't, that could put gigantic forces on the riders. Remember, as you go faster and faster, the curves have to get bigger and bigger. Banking of the coaster track is really important.

In our coaster rides, we call the G forces (gravity forces) the *seat force*. That is the force people feel in their seat. We try to keep the positive seat force of our rides at 3.5 Gs or less. One G is normal gravity here on Earth. On the roller coaster, positive G force is down through your seat.

Coaster rides have upward, or negative, seat forces too. That feeling is like being pulled off your seat into the air. For our rides, that force is half of 1 *G*. Some coaster enthusiasts call this feeling *air time*. We also make sure a rider won't get slammed across the side of the coaster car. To avoid that problem, we shoot for less than 1-*G* side force.

You don't want to have sustained high *G* forces on riders. We typically design our rides with less than 20 milliseconds of the higher forces. A longer time might make people get dizzy and have other unwanted feelings. But people can be their own worst enemies. On certain rides, everyone must heed posted signs. Sitting up straight with your head back against the head rest is really important. Obeying the signs will prevent neck injuries.

What about the future? Coaster rides will likely get bigger and bigger. The next big step is the 100-mile-per-hour coaster. That has to get really high—way over 300-feet high—for the coaster car to gain that much speed.

The limiting factor in the future might be money. The really big rides, by the time they're installed, can cost anywhere from $8 million to $12 million.

I think the roller coaster will remain the king of the park. It has been that way for more than 100 years now. I am sure the future will be equally as exciting as what is being built today.

In some ways, my job is very odd. You start the work by sitting down with a blank piece of paper. When the work is finished, people wait in line for two hours to experience what you've done. Then you hear them yelling, screaming, and laughing. There's a lot of satisfaction in that.

If people aren't screaming, it's not a good ride.

The latest thrills on dry ground

Roller coasters. Motion simulators. Thrill rides. Theme parks are rolling out attractions like never before to woo visitors this summer, experts say. A look at what's new:

Disneyland, Anaheim, Calif.

Indiana Jones Adventure. Take computerized cars for an adventure through the Temple of the Forbidden Eye. Now open.

Paramount's Great America, Santa Clara, Calif.

Nickelodeon Splat City. Kids get messy at a live game show and play areas in a three-acre themed land based on the network's shows. Now open.

(Similar attractions are open at Paramount's Kings Dominion in Doswell, Va., and Paramount's Kings Island near Cincinnati.)

Dollywood, Pigeon Forge, Tenn.

Jukebox Junction. Return to the '50s in this seven-acre themed area with a rock 'n' roll show, a diner, a '50s car ride and more. Now open.

Busch Gardens Williamsburg, Williamsburg, Va.

Escape From Pompeii. Water ride through flooded ruins ends with Vesuvius erupting. Now open.

Dorney Park & Wildwater Kingdom, Allentown, Pa.

Berenstain Bear Country. Meet Papa, Mama, Brother and Sister Bear at this children's play land, for ages 3 to 8. Now open.

Six Flags AstroWorld, Houston

Mayan Mindbender. Explore an ancient temple that holds Texas' first indoor roller coaster. Opens Saturday.

Sea World of Florida, Orlando

Wild Arctic. Ride a motion-simulator helicopter to the Arctic, then emerge in a wrecked ship surrounded by live polar bears, beluga and seals. Opens May 24.

Opryland, Nashville

The Hangman. Legs dangle on this inverted, looping roller coaster, which rises 10 stories and flips four times. Opens June 1.

Knott's Berry Farm, Buena Park, Calif., near Los Angeles

Jaguar! Enter the mysterious five-story Temple of the Jaguar and onto a sleek roller coaster. Opens June 17.

Six Flags Great Adventure, Jackson, N.J.

Viper. Walk into a Southwestern ghost town and onto a coiled coaster, where you'll get the sensation of weightlessness with a 360-degree spin. Opens in June.

Magic Kingdom, Lake Buena Vista, Fla.

ExtraTERRORestrial Alien Encounter. What appears to be a theater show becomes a thrill attraction as special effects create the illusion of a monster on the loose. Opens late June.

Universal Studios Florida, Orlando

A Day in the Park With Barney. Sing along with Barney at a musical show, then move on to a hands-on educational play area themed to his show. Opens in July.

— *Gene Sloan*

USA TODAY, 12 MAY, 1995

Ferris Wheels, Figure Eights, and Moon Trips

America's industrial revolution brought with it wider use of gears, iron, steel, and electricity. Technology helped advance thrill rides.

Using the latest engineering techniques, George Washington Gale Ferris erected his Ferris Wheel in 1893. The gigantic wheel, 264 feet in diameter, was built for the World's Columbian Exposition in Chicago. The grand structure weighed 1,200 tons and could give an uplifting experience to 2,160 riders at once.

At the turn of the century, the first figure-eight roller coasters began running. This type of coaster track weaves over and under itself.

The 1901 Pan-American Exposition in Buffalo, New York, showcased one of the first illusion rides. The attraction was really out of this world! For the Trip to the Moon, some 30 passengers boarded a structure similar to an airship. Attendants then rocked the ship back and forth, simulating a blast-off for the Moon. Looking out portholes, riders flew over the fairgrounds and Niagara Falls. Once landed on the Moon, passengers debarked and walked on a fake lunar landscape. Each rider was given a souvenir chunk of cheese for taking the voyage. The trip seemed so real, several thrill seekers passed out.

If you didn't like rocketing to the Moon, airplane swings were available. They appeared just after the Wright Brothers took to the sky in late 1903. Aerial rides became very popular as part of amusement park fare.

In 1909, designer and builder John Miller pioneered high-speed coasters and numerous safety devices. He has been called the Thomas Edison of roller coasters. Miller's patents included innovative designs for brakes, coaster cars, and wheels. Using his designs, megacoaster rides with vicious curves and deep dips became feasible.

Thrill seekers were treated to many forces on roller coasters. Any good high-speed coaster ride could offer short bursts of force, from severe gravitational stress to complete weightlessness. Albert Einstein would later call the roller coaster a perfect example of energy conversion in a mechanical system.

Giant Wheel at Cedar Point in Sandusky, Ohio at night.

Roller coasters were on a roll. The rush was on to find new ways safely to speed, lift, drop, spin, jolt, whirl, and otherwise frighten a customer, all for the price of a ticket. The concept was simple: People will plunk down good money to be scared out of their wits. For some, a death-defying ride was not enough. Among white-knuckled riders holding on for their lives, there was always a showoff or two. These were the "Look, Ma, no hands" daredevils. Bars, belts, and other restraints today make it more difficult to be a daredevil.

By 1929, recreational parks across the United States had nearly 1,300 roller coaster structures. Then America was engulfed in the Great Depression and the start of World War II. Extra cash and the public's appetite for entertainment disappeared. The scream machines fell silent.

Give It a Whirl

Purpose

To investigate the forces exerted on a person riding on a revolving ride.

Background

You work for an amusement park ride design company. The company has asked you to design a new ride that has a seat attached to the end of a cable. The cable and seat should be able to extend outward as the ride revolves faster and faster. Build the ride so it demonstrates centripetal force and *G* force. A coworker has already designed a model that shows both. Your task is to incorporate the design into a new and different thrill ride.

Materials

For each group:

- Goggles for each group member
- Plastic soda straw
- Meterstick
- 1 meter of string
- 3 large paper clips
- 10 small washers (all the same size)
- Clock or watch
- Several sheets of blank paper
- Markers or crayons

Procedure

1. Thread the string through a straw and tie a paper clip to each end of the string.
2. To simulate a rider, attach a washer to one of the clips. Attach two washers to the other clip. These two washers represent 2 *G*s of centripetal force on the rider.

Wave Swinger at Cedar Point in Sandusky, Ohio.

3. Pull the rider end of the string so it extends 40 centimeters from the straw to the rider. Hold the rider at that distance while a partner attaches a paper clip to the string where it comes out of the other end of the straw. Do not tie it, because it has to move freely! This is the marker.
4. Hold the straw vertically with the two washers hanging down. Let go of the rider and record what happens.
5. Now start the rider whirling in a circle parallel to the floor. Increase the rate of spin until the marker clip is pulled up against the bottom end of the straw. Keep swinging the rider around so the marker clip spins freely just below the straw.

CAUTION: Keep clear of others while swinging the washer.

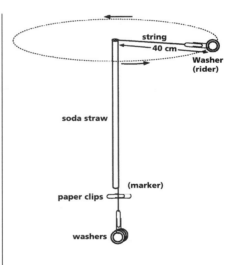

6. As one partner swings the rider around, another should count the number of circles made in 30 seconds. Make this count three times and take an average of the counts. Multiply the average by two to get the number of revolutions per minute (RPM) for the washer.

7. Sketch your contraption, estimating the angle formed between the straw and the rider.
8. Repeat the experiment several times with additional *G*s attached. Record your observations and draw diagrams of what you see.
9. Decide on a constant number of *G*s and run a set of trials using other distances between straw and rider. Try at least one distance greater than 40 centimeters and one less than 40 centimeters.
10. Display your results in a creative way.

Conclusion

The company you work for wants you to submit your whirling ride concept in the form of a diagram. What will it actually look like? They also want answers to several questions (see below).

Prepare a color drawing of a ride based on your experiments and prepare a memo to your company president with the answer to these questions:

1. What happened to the RPMs of the rider when you added additional *G*s?
2. What happened to the RPMs of the rider when you decreased the radius of the circular path?
3. As the number of *G*s increased, how did the angle between the straw and rider change?
4. The revolving rider is constantly changing its direction of motion.

Newton's second law says a change in motion must be caused by a force. What produces the force that causes the circular motion of the rider? What happens to the motion of the rider as this force is increased?

5. According to Newton's third law, the rider must exert an equal and opposite force on the string. How did this experiment show this equal and opposite reaction?
6. Predict what would happen if the string attached to the rider suddenly broke. In what direction will the rider go?
7. What new things did you learn about the design of amusement park rides from this activity?

Soak City at Cedar Point in Sandusky, Ohio.

The Force Is With You

Watch a roller coaster car speeding down its first hill. Newton's first law of motion says the car will continue to move in the direction it is already heading, unless an outside force changes its direction.

When the speeding coaster car comes to a curved section of track, it does not follow a straight path. Instead, it swiftly curves to follow the track. A force must be acting on the car! Otherwise, the car would continue in a straight line.

The force acting on the car is called *centripetal force*. Centripetal force is the amusement park's true friend. *Centripetal* means "toward the center." Where does this force come from? What gives the car its push toward the center of the circle?

The track itself does the pushing. Curved sections of track push the wheels, forcing them to follow the curve. Sometimes the curved tracks are banked, too. That is, the outer track is raised higher than the inner track. The banking, when combined with the curve of the track, allows the coaster to make even sharper turns.

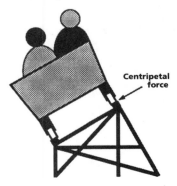

Centripetal force

Magnum XL-200 roller coaster at Cedar Point in Sandusky, Ohio.

You might have heard of *centrifugal force* (force "away from the center"). It is the apparent force that seems to push you against the outside wall of the coaster car when the roller coaster whips around a curve. This thrust toward the outside might be called centrifugal force, but it's not really a force. Nothing is pushing or pulling on you.

Your body, moving at the same speed as the coaster car, is traveling straight ahead at every point on the curve. Newton's first law of motion is acting on your body. Inertia makes your body move in a straight line. Centripetal force, applied by the track, is acting on the car. You are strapped in and hang on. The car transmits the centripetal force to you.

Instead of continuing along your straight path, you curve with the coaster car. It feels as though something is pulling you

against the person next to you or against the outside wall of the car. That is your own inertia resisting the change in direction.

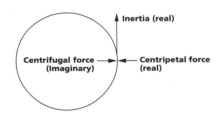

Inertia (real)

Centrifugal force
(Imaginary)

Centripetal force
(real)

Think about being in a ride that is a rotating, circular container (often called a *centrifuge*). As the ride spins, everything—including you—has inertia. Everything wants to move in a straight line; but everything is being pushed toward the center of rotation by centripetal force.

Future space colonies might be large rotating structures. Inside, people will be able to live comfortably because the speed of rotation will be adjusted so the centripetal force is

equal to 1 *G*—the force of gravity here on Earth. For space colonists, "down" will be created by the body's reaction to centripetal force.

The Shock Wave at Six Flags Great America park in Gurnee, Illinois, is a centripetal thrill ride. It uses a design called a *Klothoid loop*—a curve that is higher than it is wide. This shape was first described by the eighteenth-century Swiss mathematician Leonhard Euler. It is a perfect shape for the roller coaster somersault.

The Klothoid loop is actually not a loop. It has a teardrop shape. You would think a 360-degree circle would be ideal, but looping a circle puts too much force on a car passenger—too many *G*s.

The Klothoid loop is less stressful. It has a radius that decreases along the upward part of its shape. The continual decrease in radius means greater centripetal force at lower speeds. When coaster cars are upside down, the combination of centripetal force, speed, and inertia keeps passengers in their seats.

On your next loop-the-loop ride, don't be too scared. Remember, the force is with you.

Magnum XL-200 roller coaster at Cedar Point in Sandusky, Ohio.

Gravity of the Situation

Here on Earth, we live with gravity every day. We depend on it to put the thrill in our thrill rides and to get the ketchup out of the bottle. However, it sometimes works against us. For example, it makes things fall to the floor and break.

More than 300 years ago, Sir Isaac Newton described the true nature of gravity for the first time.

Gravity, he said, is a force throughout the universe. Every object pulls every other object toward itself by exerting a gravitational force. The more massive an object is, the greater its gravity. The farther away an object is, the weaker its gravity.

Raptor at Cedar Point in Sandusky, Ohio.

Magnum XL-200 at Cedar Point in Sandusky, Ohio.

One day, an apple fell from a tree and struck Newton on the head. According to the story, he realized that the same force that made the apple fall extends all the way to the Moon, holding it in orbit around us.

In the same way, the gravity of the Sun and other planets influences the orbit of each planet. Because the Sun has a much greater mass, it influences the planets far more than they influence each other or the Sun.

Newton's law of universal gravitation is a mathematical statement. It says that the attractive force between any two objects increases in proportion to the product of the two masses. It also says that the attractive force decreases in proportion to the square of the distance separating them.

If every object attracts every other object, why don't you feel an attraction as you walk past a big building? After all, the building's mass is much greater than yours. The answer is that Earth is by far the most massive object near you. It exerts the mightiest force on us all. The mass of any object on Earth is small compared to Earth's mass, and all objects on Earth are strongly held by Earth's gravity.

Sensitive scientific instruments can measure the small gravitational attraction of big buildings or great mountains by subtracting Earth's very large gravitational pull from their measurements.

The Swing of Things

Purpose

To study forces that influence the motion of a pendulum, and to identify energy transformations in pendulums.

Background

You are an amusement park operator with a pendulum ride. All it does is swing people back and forth. The ride was very popular when it first opened. Long lines formed as people anxiously awaited their chance to swing. Now, the pendulum ride has lost its appeal. You are losing money staffing it, and you must do something to make it interesting again.

Some of the complaints from riders are, "It doesn't last long enough," and "It's not very exciting. I want a real thrill out of this ride." Before you meet with the design engineer, you want to learn more about pendulum motion so that you will not sound like a swinging idiot during the meeting.

Materials

For each pair:
- About 150 cm of string
- Scissors
- Pencil
- Masking tape
- Large paper clip
- 5 or more washers
- Meterstick
- Clock or watch

Procedure

1. Tape a pencil to the edge of a desk or table so the eraser part extends beyond the edge. If other supports are available, use them too. The pencil acts as the pendulum's pivot point.

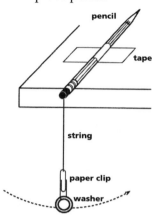

2. Measure about 50 centimeters of string and tie a paper clip to one end. Slip one washer on the paper clip. Attach the other end of the string to the pencil. Carefully measure and record the length of the string from the pencil to the center of the washer.
3. Pull the washer to the side about 20 centimeters and let it swing. Count the number of full swings (back and forth) in one minute. Repeat the procedure and find the average. Record this average number of swings in a data table. Also describe what happens to the size of each swing (its *amplitude*) as time passes. This first set of trials represents the unpopular ride you want to change.
4. What variables (such as string length) can you change to test different pendulum designs? Conduct a series of experiments, changing one variable at a time. Each time you run an experiment, record the number of swings in one minute. Record what happens to the amplitude after one minute. Present the data in an organized form.

Conclusion

Making the swing ride more interesting will cost money. But as they say, it takes money to make money. Prepare a memo to the owners of the amusement park. In the memo, state your recommendation for making the swing ride more interesting. Estimate the cost of the changes required (any reasonable estimate will be accepted). In addition, you must explain the science behind your recommendation.

Be sure to explain the science in simple terms. The owners of the amusement park are scientifically illiterate.

Here are some questions to consider as you write your memo.

1. What force keeps a pendulum going? What forces cause a pendulum to come to a stop?
2. At what point in a pendulum's swing does the washer have the most potential energy? At what point does it have the most kinetic energy?
3. What is happening to the total energy of the pendulum as time goes on?
4. Will lifting the ride higher at the start produce a longer ride or faster ride?

Virtual reality lessons hit home

High-tech vans bring kids hands-on experience

By Belinda Thurston
USA TODAY

Rainfall and evaporation.

Water comes down. Then it goes up. No big deal, right?

But through the eyes of the molecules involved, it's major action.

And that's just the perspective students from 14 Western states are getting through virtual reality.

Three vans stocked with Space Age-looking hoods and computers, called Virtual Reality Roving Vehicles, began touring Washington schools in November. The $1.25 million project, run by Seattle's Human Interface Technology Lab, affiliated with the University of Washington and the U.S. West foundation, demonstrates virtual reality technology to elementary, middle- and high-school students.

Virtual reality is a 3-D, computer-generated environment in which the computer user can make things happen. By using a visor and hand controls, students can actually enter a microscopic world and construct a water molecule, picking up electrons and neutrons as if they were tennis balls.

The three vans tour public and private schools in Washington, Arizona, Colorado, Idaho, Iowa, Minnesota, Montana, Nebras-

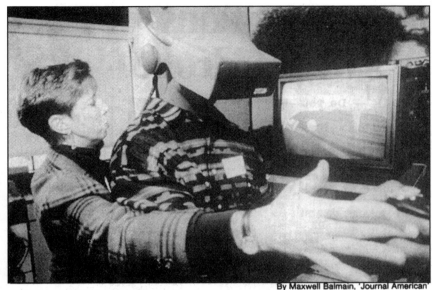

By Maxwell Balmain, 'Journal American'
LEARNING VIRTUALLY: Kimberley Osberg, operations manager of Virtual Reality Roving Vehicles, helps Seattle senior Karl Smith explore the basics of chemistry.

ka, New Mexico, North Dakota, Oregon, South Dakota, Utah and Wyoming through June.

Kimberley Osberg, operations manager of the project, says the goal is to expose students to new technology as well as study whether it helps them learn.

"Instead of talking through what you want to convey, you're showing, feeling and doing," Osberg says.

Chemistry teacher Mark Stewart, Garfield High School, Seattle, says virtual reality is just another teaching tool to reach kids with a variety of learning styles.

"In chemistry we talk about a lot of abstract ideas," Stewart says. "This may be that one tool that helps that one kid who works better hands-on."

USA TODAY, 11 APRIL, 1995

My favorite ride is the Anaconda. It's not too rough, and it's not too big, it's just fun. You're pushed in every direction, and when you're upside down, it's really cool because you feel like you're going to fall out.

I don't like wooden rides very much because I feel that if a board broke, it could be disastrous. I think the metal rides are safer.

DAWN COMPTON
RIXEYVILLE, VA

You've Got Potential

The slippery trough twists to the left just below your feet. You feel a spray of cool water as you slide through the curve. You are gaining speed with every turn. It is a long way down, but it is over too quickly.

When you sit at the top of a water slide, you are a storehouse of energy. You have *potential energy*—the stored energy of your position above the ground.

As you descend, you are obeying a physical law—the law of conservation of energy. The law says energy can change from one form to another, but energy cannot be created or destroyed. All of the energy in the universe remains constant.

As you start your slide, some of your potential energy begins to change into another form of energy called kinetic energy. *Kinetic energy* is the energy of motion. The potential energy you had before you started to slide is transformed into kinetic energy as gravity pulls you lower.

Now that you are wet, climb an even taller slide. The higher climb is worth the extra effort because you now have even more potential energy.

Once again, as you swish down, your potential energy changes to kinetic energy. The higher slide allows you to go faster and farther. Because you started with more potential en-

Corkscrew at Cedar Point in Sandusky, Ohio.

ergy, you wind up with more kinetic energy as you scream off the end of the slide.

While you are thinking about energy, where did you get the energy to climb to the top of the slide? What happened to your kinetic energy when you hit the water at the bottom of the slide?

Blue Streak at Cedar Point in Sandusky, Ohio.

Screaming for Speed

Clack, clack, clack—gears slowly pull the roller coaster cars to the top of the first hill. The train of cars arches as it passes over the top. Then the cars start their descent. They move slowly at first, but rapidly gain speed. Faster and faster they fly. Long before they reach the dip and head up the next hill, all the passengers are screaming.

The cars slow slightly going up the second hill. Shrieking passengers momentarily lift from their seats at the hill's crest. The cars streak down the second slope, accelerating, gaining the momentum that will carry them over the third hill.

Engineers design the slopes of gravity thrill rides so acceleration going downhill is greater than deceleration going uphill. That keeps the coaster cars moving without having to add any energy to the system. To stop the cars, engineers use mechanical brakes.

What are *acceleration* and *deceleration*? The words mean "speeding up" and "slowing down." The root Latin word, *celer*, means "swift."

Velocity is the speed of an object traveling in a specific direction. Acceleration is a measure of how fast an object's velocity is changing.

The units of acceleration are a combination of units of speed and units of time—meters per second per second, or meters per second squared.

When the velocity of an object increases, the object is accel-

Raptor, inverted roller coaster at Cedar Point in Sandusky, Ohio.

erating. When the velocity decreases, the object is decelerating. A roller coaster moving at a constant speed of 50 miles per hour has an acceleration of 0.

A coaster headed downhill is almost a freely falling object. The force accelerating it is gravity. Free-falling objects accelerate at the constant rate of 9.8 meters per second squared. The coaster accelerates at a constant rate that is almost that great.

From a standing start with a velocity of 0, a freely falling coaster reaches a speed of 9.8 meters per second (22 miles per hour) after the first second, 19.6

meters per second after two seconds (44 miles per hour), and 29.4 meters per second after three seconds (65 miles per hour). Its actual speed at the bottom of a slope depends on the height from which it started and the amount of slope.

The Italian scientist Galileo Galilei was first to state the law of constant acceleration. From experiments he did in 1589 with balls rolling down inclines, he proposed that, in the absence of forces other than gravity, all objects will fall at the same rate despite their masses. There is a catch, however. You need to take air resistance away.

Galileo's finding was demonstrated nearly 400 years later on the Moon, where there is no air. On August 2, 1971, Apollo 15 astronaut David R. Scott dropped a feather and hammer in the Moon's vacuum. Released at the same time, both landed on the lunar surface at the same time.

Thrill rides deliver acceleration in different ways:

- A ride moving downhill in a straight line speeds up. The direction of the acceleration is in the direction of motion.
- A ride moving uphill in a straight line slows down. The direction of acceleration is opposite the direction of motion. This is deceleration.
- A ride moving in a circle at a constant speed is also accelerating. The direction of acceleration is toward the center of the circle.
- A ride speeding over a hill, which is a parabolic curve, is accelerating. The direction of acceleration is along the axis of the parabola.

Several directions of acceleration can affect a roller coaster rider:

- *Vertical acceleration* is perpendicular, or at a right angle (90 degrees), to the track and toward it. You feel compressed in your seat. The greater the acceleration, the more squashed you feel.
- *Longitudinal acceleration* is in the direction of the coaster's motion. You feel pushed back against your seat. Your head and shoulders may swing backward.
- *Lateral acceleration* is to the side, relative to the coaster's motion. It makes you slide sideways across the seat. You might even squash your ride partner.

In an accelerating coaster, potential energy is changing to kinetic energy. A coaster decelerates when it loses kinetic energy. Twists and turns in a track help decrease the kinetic energy. Friction between wheels and rails slows the cars. Air resistance, or drag (the friction between the cars and the air), also plays a part. Brakes bring the cars to a stop, ultimately dissipating any remaining kinetic energy in the friction between brakes and wheels.

The passengers stop screaming, get out of the cars, and line up to buy tickets for another ride.

It Sounds Like Work to Me

According to scientists, math homework is not work. Science homework is not work either. Then what is work?

To a scientist, *work* has a very precise meaning. Work is done when a force acts on a mass and causes it to move. The amount of work done is the product of the force exerted times the distance moved.

work = force x distance

Energy is related to work because it takes energy to do work. Units of energy are measures of the amount of work done. The joule is the unit of energy in the metric system. One joule is the amount of energy needed to produce 1 newton of force over a distance of 1 meter.

Energy may be electrical, mechanical, chemical, thermal, or nuclear. It can be used now or stored for later use. Stored energy (potential energy) has the capacity to do its work in the future. Kinetic energy is doing its work now.

A tightly stretched rubber band has potential mechanical energy. Can you think of other examples of stored energy?

So Nice of You to Drop In

Purpose
To apply Newton's second law to the motion of falling objects, and to design and demonstrate a parachute-drop device.

Background
A famous amusement park company has hired you and your partner(s) as consultants. They need your help designing a parachute ride. Their attorneys are worried. A parachute that drops too fast might injure riders. The parachute must also land the riders in a small area directly under the release point. Otherwise, people might crash onto the shaved ice stand.

Before you draft your reply, you have a few days to use the laboratory to sharpen your understanding of falling objects and parachute design.

Materials
For each group:
- Paper clip
- 5 washers
- Several meters of string
- Scissors
- Stopwatch
- Hoist mechanism or cardboard tube with extra string (optional)
- Several sheets of blank paper
- Tape

Procedure
1. Slide five washers onto a paper clip, then devise a way to drop this mass from a high place in the classroom. The drop height should be a minimum of 3 meters. Measure the drop height to the nearest hundredth of a meter.
2. Use a stopwatch to measure the free-fall drop time for the mass to the nearest tenth of a second. Repeat for a total of five trials. Create a table to record the data. Record times in nearest tenths of seconds. Calculate and record the average free-fall time.
3. Use the chart to find a calculated drop height based on your average free-fall time. If time permits, collect data for other heights.

Free-Fall Time (s)	Drop Ht (m)	Free-Fall Time (s)	Drop Ht (m)	Free-Fall Time (s)	Drop Ht (m)	Free-Fall Time (s)	Drop Ht (m)
0.05	0.01	1.05	5.40	2.05	20.59	3.05	45.58
0.10	0.05	1.10	5.93	2.10	21.61	3.10	47.09
0.15	0.11	1.15	6.48	2.15	22.65	3.15	48.62
0.20	0.20	1.20	7.06	2.20	23.72	3.20	50.18
0.25	0.31	1.25	7.66	2.25	24.81	3.25	51.76
0.30	0.44	1.30	8.28	2.30	25.92	3.30	53.36
0.35	0.60	1.35	8.93	2.35	27.06	3.35	54.99
0.40	0.78	1.40	9.60	2.40	28.22	3.40	56.64
0.45	0.99	1.45	10.30	2.45	29.41	3.45	58.32
0.50	1.23	1.50	11.02	2.50	30.62	3.50	60.02
0.55	1.48	1.55	11.77	2.55	31.86	3.55	61.75
0.60	1.76	1.60	12.54	2.60	33.12	3.60	63.50
0.65	2.07	1.65	13.34	2.65	34.41	3.65	65.28
0.70	2.40	1.70	14.16	2.70	35.72	3.70	67.08
0.75	2.76	1.75	15.01	2.75	37.06	3.75	68.91
0.80	3.14	1.80	15.88	2.80	38.42	3.80	70.76
0.85	3.54	1.85	16.77	2.85	39.80	3.85	72.63
0.90	3.97	1.90	17.69	2.90	41.21	3.90	74.53
0.95	4.42	1.95	18.63	2.95	42.64	3.95	76.45
1.00	4.90	2.00	19.60	3.00	44.10	4.00	78.40

4. Make a parachute out of paper and fasten it to the five-washer mass using short lengths of string. The parachute can be any shape.

5. Tape a small paper loop to the top of the chute so you can hold the mass at the drop height. Start the mass under the parachute at the same drop height as you did for the free-fall trials.

6. Let go of the parachute. Find and record the parachute drop time. Repeat for five trials, and calculate the average parachute drop time. Organize your data into a table.

7. Now perfect your parachute. You may redesign it in any way, but your goal is to have the drop time as long as possible and the landing as accurate as possible (directly under the drop point). Demonstrate your design for the class.

8. When all groups have demonstrated their parachute drops, brainstorm as a class why some designs worked better than others. Take notes during the discussion of points a–f below. The answers will help you with your response to the amusement park company.

a. Compare the height you got from the chart provided with the measured drop height of the object. If the two heights are not equal, suggest reasons why they are different.

b. What do you think would happen to the free-fall time if you dropped a more massive object from the same height? How about something with less mass? Speculate, then experiment to test your ideas.

c. Release a mass to fall freely to the floor. Compare the mass's speed just after release with its speed near the floor. Make the same comparison for a mass with a parachute attached to it.

d. According to Newton's second law, any acceleration is caused by a force or a group of forces acting together. Compare the acceleration of the free-fall mass to the mass on the parachute.

e. Discuss what other force or forces might be canceling

the effect of gravity when you use a parachute to drop a mass.

f. Your challenge was to design a parachute that causes the mass to drop in the longest time while landing directly below the drop point. From the brainstorming you did with your colleagues, what things made some designs work better than others? Which was more difficult, slowing the drop time or improving the landing accuracy?

Conclusion

Draft your reply to the company that hired you. You may use an informal fax format, or the more formal memo format. Be sure to address their concerns. As you prepare your response, pay particular attention to the answers to the six questions.

Three Laws for the Price of One

Do you want to experience all three of Newton's laws of motion for the price of one ride? Try bumper cars!

What happens when you run your bumper car into another person's bumper car? Your car stops, but your body lurches forward. That's inertia at work—Newton's first law. Your body continues in motion until it encounters a force to stop it.

If you drive slowly and lightly tap another car, the other car only moves a little bit. At higher speeds, your moving bumper car packs a real wallop. You become a force in motion! When you hit another car, the force of the hit causes it to accelerate away from you. These collisions demonstrate the relationships among force, mass, and acceleration—Newton's second law.

When your bumper car smacks into a bumper car that is just sitting there, your car exerts a force on the sitting car, sending it moving. The sitting car exerts an equal force on your car that sends it moving, too. This is Newton's third law—action and reaction.

On your next bumper car ride, enjoy your experience with

Disney magic at work in Europe

By Martha T. Moore
USA TODAY

Walt Disney's sleeping beauty is finally waking up. Euro Disney, the troubled theme park complex outside Paris, reported a profit Tuesday for the first time since opening in 1992.

That's good news for Disney, which holds a 39% stake in Euro Disney and reports its quarterly earnings today. They're expected to be about 55 cents a share, up from 49 cents a year ago.

Boosted by higher attendance and a new ride, Space Mountain, Euro Disney reported a quarterly profit of 170 million French francs ($35.3 million) the second quarter, vs. a loss of 546 million francs ($113.5 million) the same period last year. Revenue was up 17% to $283 million.

The news sent Euro Disney's stock up 1.15 francs (24 cents)

COMPANY SPOTLIGHT
A DAILY LOOK AT A COMPANY, INDUSTRY OR MARKET TREND

to 17.45 francs ($3.63) in Paris. In New York, Disney shares closed at $55¾, up ¾.

"We certainly have a chance to break even in 1995," a year ahead of schedule, says Chief Operating Officer Steve Burke. Analyst Marc St. John Webb of ING Bourse in Paris predicts Euro Disney could earn $31 million for the year. Less aggressive analysts predict a $10.4 million profit.

To make it into the black, Euro Disney has overhauled virtually every aspect of its operation, including its name. The theme park is now called Disneyland Paris. Hotel, restaurant and merchandise prices have been cut, and entrance prices were slashed 22% in April, to $41 from $52 for a one-day adult ticket. The company also cut costs by

about $100 million last year. And it has added 10 attractions. But the company is still struggling to get visitors to spend more on food and souvenirs. And plans for a second, adjoining theme park — which would improve revenue by lengthening the average visit — are still up in the air. "We need to put a few more good quarters together before we would be in a position to take that off hold," Burke says.

In a financial restructuring last summer, Euro Disney's banks agreed to suspend interest payments on the company's $3.1 billion debt until 1998. Disney also will not collect royalties or management fees until then.

Euro Disney's results are unlikely to affect Disney's earnings, analysts say. "When

Product sales key to earnings

Walt Disney (DIS)	'93	'94	'95 est.	'96 est.
Revenue (billions)	$8.5	$10.1	$11.7	$12.3
Net income (billions)	$0.30	$1.11	$1.37	$1.60
Earnings per share	$0.55	$2.04	$2.60	$3.00
Hq: Burbank, Calif.		Exch: NYSE	Employees: 65,000	
Div./yld.: $0.36 / 0.65%		P-E: 19	Shares: 521 million	
52-week high/low: $60 / $37¾			Tues. price: $55¾, +¾	

Earnings, revenue estimates: Merrill Lynch
P-E based on estimated 1996 earnings.
Source: USA TODAY research

you've got 530 million shares, it takes a lot of performance at Euro Disney to move (Disney stock) more than a penny or two," says Disney analyst Jeffrey Logsdon of The Seidler Cos.

More likely to boost Disney's earnings are higher theme park attendance and strong consumer product sales. Revenue from films is expected to be weaker, analysts say, compared with 1994's successful Return of Jafar video. "We will have Pocahontas (vs.) The Lion King issues, but that's a bigger deal in September," when this quarter's earnings are released, says analyst Sharon Williams of C.J. Lawrence Deutsche Bank Securities.

USA TODAY, 26 JULY, 1995

Physicist

ROBERT SPEERS
FIRELANDS COLLEGE
BOWLING GREEN STATE
UNIVERSITY
HURON, OHIO

I'm an associate professor of physics at Firelands College. I use roller coasters and other rides to help explain physics to students and teachers.

Near the college is a famous amusement park called Cedar Point. It's located in Sandusky, Ohio. I'm no stranger to the rides there. As a kid, I lived across the bay from the park. Cedar Point has been an enjoyment for me for many, many years.

Every couple of years, a major new ride is installed there. Each year, the rides seem bigger, better, and different. A recent addition is the tallest, fastest, and steepest standup roller coaster!

My interest in physics began in the mid-1950s. I wanted to find out why the world worked the way it did, in a general sense. The space race between the United States and the then Soviet Union was very strong in those years. Rocketry and space seemed like a very interesting area as a career.

I was intrigued by the questions the physics people were asking at the time. What was the Moon really like? What was the Moon made of? Is gravity different there than here? What would it be like up in orbit? How

will plants grow in space? There were a lot of questions.

Along with an interest in engineering, I found the atomic and nuclear side of things fascinating. Nuclear reactors were being built. There were lots of unanswered questions.

In high school, I can't say I was a top student. But I did dedicate my interests to one area and kept after it. I took physics in undergraduate school.

In the 1960s, I started working in the semiconductor field. Those are electronic components now in use everywhere. Integrating semiconductors into circuits was just a vision somebody had back then.

Then I attended Ohio State University. My research there involved studying solid-state and semiconductor electronics. I received my master's degree and then a doctorate in solid-state electronics. From there, I began to work for RCA, a research cor-

poration in Princeton, New Jersey. Developing infrared television systems was my job then.

Following that experience, teaching seemed like an attractive thing to do with my life. Firelands College was being built next to my hometown, so I came here and shifted back into pure physics. It's been a nice journey learning about science, physics, and now teaching.

My teaching at the college begins early in the day. There's lots to do working with students in the laboratories. I set up the lab experiments and repair and buy equipment. These days, computers are used quite a bit in lab work. So a constant thing is to keep the computers running properly. I'm always installing new software or hardware.

Other parts of my day involve grading papers. My contact with students is not just at school, but in outside tutoring as well.

If you're interested in pursuing a career in physics, doing well in math is important. Take college prep math classes. Courses in the sciences, such as biology, chemistry, and physics, are important. Environmental science classes are good to take as well. Take shop courses, too. They turned out to be very helpful for me. Making a drawing of a gizmo means communicating your idea to others. That's very important.

Remember, there are many types of physicists. Some are theoretical, looking at the far edge of physics. Many are experimental physicists who build things. They need to understand why things work the way they do. Physicists also have careers in molecular physics, in nuclear physics, and in biomedical fields.

Here's something to remember. Many great physicists had learning disabilities. Albert Einstein and others didn't do well in the traditional course work. They had trouble focusing and were always daydreaming. They used their creative talents differently.

Using physics in my daily life …that happens when I'm driving a car. It can snow and rain quite a bit here in the northern part of the country. We've got lots of slippery roads. When I'm driving, I always think to myself: How fast should I be going? What's the right speed for me to take that curve in the road?

Even driving over a hill brings up some physics. How much does the car weigh? It's really slippery, so I'd better watch the traction. Also, the brakes won't work as well. Neither will the steering.

I drive a Mustang convertible with rear-wheel drive. It has a big V-8 motor. That car is a challenge to drive in the winter. Look in my trunk—I've got a load of bricks in there so I don't do donuts in the snow!

If you are trying to come up with a new ride, I have some thoughts. Take a hard look at examples of current rides. You might want to go to a playground and think through some different ideas. Write down your impressions of a playground ride. What would you do if there were no boundaries on safety? Let your imagination flow.

Perhaps you want to figure out the *G*-force level you want a passenger to experience. Making a two-dimensional drawing or building a three-dimensional model is helpful. That's an essential communication skill.

Don't forget about the feeling of a ride. Perhaps 99 percent of the way we interact with the world is visual. The rest is feeling. Try to ignore the visual part of things. Feel what is going on with your body during a ride.

A real pleasure for me is teaching physics at the amusement park. I organize Physics Day at Cedar Point each May. The program involves high school teachers and about 8,000 of their students.

Students have assignment sheets. We hand out instruments to measure the various forces they experience on the rides. Students record data, such as the *G* forces and acceleration forces. The instruments also record time in free fall. Students then compare their theoretical predictions about the ride with their actual measurements. That allows them to better understand the physics behind thrill rides. You might call it in-your-face physics.

If there is an amusement park near your school, perhaps your science teacher will arrange a field trip for your class. There is no better place to learn Newton's laws of motion.

G-Whiz, I'm Falling

The unit of measurement used to describe gravity forces is the *G*. On this planet, we say the gravity force we experience is a 1-*G* force. How many *G*s do you feel on thrill rides?

Roller coasters are sometimes called gravity rides. Machinery pulls the coaster cars to the top of the first hill. As the cars reach the crest, they are released and gravity takes over. From that point, gravity does all the work. At the very top of the first hill, before starting that first plunge, the coaster cars have their maximum potential energy.

During sharp twists and turns of a coaster ride, centripetal forces come into play. The rider feels heavy at the bottom of a long drop or rounding a corner. The higher the speed and sharper the turn, the higher the *G* force experienced.

For example, the Steel Phantom at Kennywood Park in West Mifflin, Pennsylvania, has the world's longest hill drop—70 meters (about 225 feet). Coaster cars there reach a speed of 130 kilometers (about 80 miles) per hour. Passengers then endure four inversions, a vertical loop, a boomerang, and a corkscrew. All this action takes just 1 minute, 45 seconds.

Many theme parks boast of rides that hit speeds of more than 100 kilometers (60 miles) per hour and cause passengers to experience well over 6 *G*s. That is six times Earth's normal gravity. At that moment, you would weigh six times more than you weigh at 1 *G*! That's a lot for the human body to endure.

Engineers have learned many lessons building roller coasters. One early roller coaster ride involved a wicked loop-the-loop. A rider would briefly feel 12 *G*s taking the ride. Many customers complained of backaches and neck pains. The *G* force was too great.

Jet pilots in their F-16 fighter planes can take high *G* forces because they wear special pressure suits. The suits make sure their blood stays distributed correctly through their bodies. Correct distribution prevents jet pilots from blacking out from blood loss in the brain. They can then make sharp turns and other maneuvers in the air. But taking forces as high as 11 *G*s for long periods is dangerous. Even jet fighter pilots would black out.

A roller coaster rider can feel weightless as well as heavy. Roller coaster hills are curved in parabolas. (A *parabola* is an open, bowl-shaped curve.) Just as riders go over the top and start down the slope, they enter free-fall.

This floating feeling is what an astronaut feels in space. A shuttle astronaut is constantly falling around and around Earth. During free-fall, the force you feel is less than 1 *G*. This condition is called *microgravity*.

NASA trains its astronauts for microgravity using a specially equipped airplane. You could call it the ultimate thrill ride.

The plane flies a parabolic trajectory. It climbs at a 45-degree angle, then dives at a 45-degree angle. As the plane arcs over the high point of the parabola, everything in the plane is in microgravity—free-falling for about 60 seconds. Bringing the plane out of the dive produces a 2.5-*G* force.

Movie makers used a plane to achieve free-fall for the film *Apollo 13*. Many scenes in the movie were filmed in free-fall. The actors felt short periods of microgravity without actually going into space.

On your next coaster ride, enjoy the gravity of your situation—the heaviness you feel at the bottom of each hill and the weightlessness as you free-fall!

Theme parks bring movie action to life

By Gene Sloan
USA TODAY

The best place to see the movies this summer might not be at a theater at all. Try the local theme park instead.

Just about every summer blockbuster, it seems, also is

appearing at a park somewhere — either as a live stage show or a "behind-the-scenes" attraction.

Reason: More parks nowadays are owned by movie companies. "Synergy," says George Ladyman, vice president at Six Flags, which has a *Batman Forever* attraction produced by sister company Warner Bros. studio.

"It's a logical brand extension," he explains. "People are wanting to see the characters, the scenes, the props."

Among current and coming attractions:

▶ *Batman Forever.* The 20-minute *Batman Forever Stunt Show* pits the Caped Crusader against the Riddler and Two-Face with pyrotechnics, karate fights, motorcycle jumps and more at Six Flags Magic Mountain in Cali-

BAT SHOW: After the movie, catch the 'Batman Forever' stunt show at Six Flags, starring Dr. Chase Meridian, Robin, Batman, Riddler and his henchman and, of course, the Batmobile.

fornia and Six Flags Great Adventure in New Jersey. The comic-book hero also battles for justice in the *Batman Forever Fireworks and Laser Show,* new at Six Flags parks in California, Texas and Georgia.

▶ *Casper. Casper — Behind the Screams* offers behind-the-scenes secrets of the

Universal movie in a replica of Whipstaff Manor, the film's haunted mansion. At Universal Studios Hollywood through Labor Day. Universal Studios Florida, Orlando, has a different behind-the-scenes exhibit, *Casper on Location,* using props and special effects.

▶ *Pocahontas.* The 30-min-

ute *Spirit of Pocahontas* stage show, based on the Disney movie, recounts the love story between the Native American princess and Virginia colonist Capt. John Smith. At Disney World near Orlando and Disneyland in Anaheim, Calif.

▶ *Apollo 13. The Making of Apollo 13* offers a behind-

the-scenes look at the Universal film, including props such as a three-story lunar space module and a cut-away from the actual spacecraft's interior. At Universal Studios Hollywood through Labor Day. Universal Studios Florida has a similar show.

▶ *Waterworld.* Jim Timon, vice president at Universal Studios Hollywood, says the park has been working for more than a year on a new live-action *Waterworld* stunt show, based on the Kevin Costner movie scheduled for release July 28. The park attraction should open soon after, he says.

Rides and other attractions based on big summer movies, of course, are nothing new at theme parks. Universal Studios Hollywood, for instance, has long had a *Jaws* attraction based on the 1975 blockbuster.

But this summer there's more such attractions and they're arriving almost simultaneously with the movies they're based on in one big media splash.

"It's definitely the trend in theme parks," says Timon. "We're a multimedia society. ... You see the film, live the film (in a ride) and see how it was made" — all in the same summer.

USA TODAY, 27 JULY, 1995

I like the Grizzly because it shakes a lot and I like its tunnel and hills and the wind in my face.

Coasters should have wooden track, steep hills, lots of turns, and dark tunnels. They also should go forward and backward.

Steel roller coasters have a smoother ride, but since wooden coasters shake a lot, they seem faster. You feel more secure in the rides that you stand up in. Sitting down with a lap bar doesn't seem as safe to me.

CRAIG ROGERS
CULPEPER, VA

Imagineering the Future

In the summer of 1954, construction workers uprooted orange trees on 160 acres in Anaheim, California. Few realized that a new kind of amusement park was being planted. Disneyland opened July 17, 1955. It was the first completely planned theme park.

Walt Disney and his team of "imagineers" created a unique form of amusement park. They built a central plaza surrounded by five different worlds: Main Street U.S.A., Frontierland, Adventureland, Fantasyland, and Tomorrowland. The Jungle Cruise, Submarine Voyage, Mr. Toad's Wild Ride, and even a new-style Trip to the Moon, were highly visited rides. The latest in electronics and other technologies made these adventures possible.

Disneyland put a new spin on the roller coaster. The Matterhorn was the first steel coaster to use a tubular track. Passenger-filled cars are taken to the 146-foot summit. Let loose, the cars race through and around the icy mountain. The tobogganlike ride ends with a splash. The cars slow when they hit the water at the mountain's base.

The Magic Kingdom at Disneyland was an immediate success. The family theme park was born. Disney magic was used to design Walt Disney World in Orlando, Florida. In addition, Disney engineers constructed the Experimental Prototype Community of Tomorrow (EPCOT) in Orlando.

Many copied the Disneyland concept. The Six Flags theme parks were an early success. Their design both borrowed from the Disneyland formula and incorporated unique thrill rides. Six Flags Over Georgia, for instance, adopted the slogan "Land of Screams and Dreams."

Coaster builders designed rides that produce acceptable G-force (gravity force) levels on riders. One of the Six Flags rides is the Mind Bender, a triple-looping steel coaster ride lasting 2 minutes, 33 seconds. On the Mind Bender, the center of gravity is not at the top or bottom of the roller coaster car. It is at the rider's heart. Engineers found that by changing the shape of the loop into an elliptical corkscrew, they could minimize G forces.

Only three theme parks existed in 1970. A decade later, two dozen theme parks existed, drawing more than 60 million people. MCA, the motion picture and entertainment business giant, created two theme parks. Tourists flock to Universal Studios Hollywood in Universal City, California, and to Universal Studios Florida in Orlando.

These parks use movie themes to attract the public. Thrill rides based on movies such as King Kong, Jaws, and Earthquake: The Big One draw record crowds. Sitting in moving cars, spectators come face to face with the giant Kong or are nearly caught in the jaws of a great white shark. Cars almost tip over in the simulated rumbling quake that shakes thrill seekers. (Or are they thrill shriekers?)

Given the incredible success of theme parks, can you still find a speedy thrill ride in an old-fashioned amusement park? You bet. Go to Cedar Point near Sandusky, Ohio. More than a century old, it is one of America's longest-surviving amusement areas. Packed with thrill rides, the entertainment area is called the "Amazement Park."

Cedar Point has 56 rides including 11 roller coasters—state-of-the-art scream machines. Try the Corkscrew, the world's first triple-looping roller coaster, or the Raptor, where passengers in cars suspended from an overhead track are turned upside down six times.

Better yet, take a trip on the steel cars of Magnum XL-200. The towering roller coaster structure tops at 205 feet. Cars reach 72 miles per hour. Magnum drops passengers 194 feet at a 60-degree angle. The entire ride takes a heart-pounding 2.5 minutes. While you are at Cedar Point, put yourself on the Mean Streak. With its top speed of 65 miles per hour, it is the world's fastest wooden roller coaster.

For the fastest steel coaster, try the three-minute ride of the Desperado at Buffalo Bill's Resort and Casino near Las Vegas, Nevada. On the Desperado, you can reach a top speed of 80 miles per hour.

The longest roller coaster ride in North America uses wooden cars. It is the 7,400-foot-long track of

Giant Wheel at Cedar Point in Sandusky, Ohio.

The Beast at Paramount's Kings Island at Kings Mills, Ohio, north of Cincinnati. Total ride time is 4 minutes, 30 seconds.

What does the future hold for thrill rides? More and more, theme parks are taking on a wet look with rides that offer log flumes, wild rapids, and other water-oriented adventure. Linking amusement parks and shopping malls is another trend.

Perhaps the ultimate thrill ride is one used by astronauts and cosmonauts to prepare for the weightlessness of space. They train in large aircraft that repeatedly arc and dive through the air. The planes fly parabolic paths that create many seconds of free-fall at the crest of the curve. Private companies have begun offering these roller-coaster-like airplane rides to the public.

The late science writer Isaac Asimov envisioned amusement parks on the Moon. Can you imagine a thrill ride in just one-sixth gravity?

Today, the thrill-ride and theme-park industries face conflicting demands. More terrifying thrill rides are likely in the future. After all, what young person would miss the chance to challenge and survive a higher, faster, more daring thrill ride? However, the average age of an amusement park visitor is increasing. Older visitors may want less unnerving rides. Catering to the older ticket holder will become more important.

Theme-park experts predict more simulation rides. Sitting in a large simulator, passengers can be visually transported down waterfalls. They can fly over the highest mountains or touch down on the surface of Mars. Such rides put less physical stress on passengers. How about coupling computer-controlled, virtual-reality technology with television? That could bring the thrill ride directly into the living room.

From the fairs and celebrations of early days to the speediest of today's roller coasters, escape and entertainment seem to be a basic human desire. The thrill ride is a safe way to accept risk while confronting fear and panic. Famed coaster designer John Allen once said, "Part of the appeal is the imagined danger."

And for the price of another ticket, going on a thrill ride is easier the second time around!

Newton's Space Program

The rocket carrying the space shuttle Endeavor rises from Cape Canaveral, Florida.

A space probe whips around Earth, using the boost of Earth's gravity to head for deep space.

Orbiting the Sun, a family of worlds called planets, asteroids, moons, and comets makes up the solar system.

What do the rocket, the space probe, and the solar system members have in common? Newton explains their paths in his laws of motion and law of universal gravitation.

On its launch pad, a rocket is a system at rest. It is in balance, or *equilibrium*. Gravity pulls the resting rocket down and Earth pushes back. At blast-off, an opening permits exploding gas to escape from the bottom of the rocket. The gas rushes out at high speed, exerting an upward force that propels the rocket skyward.

The force of the rocket is greater than the force of gravity. The thrust overcomes the rocket's weight. The rocket passes from a state of rest to a state of faster and faster motion.

The rocket is obeying Newton's third law of motion. For every action, there is an equal and opposite reaction. The downward action of the expanding gases produces an upward reaction of the rocket itself.

If the rocket runs out of fuel before it is moving fast enough to escape the pull of gravity, it drifts and slows. Then it begins to fall. Gravity pulls the rocket back to Earth.

What happens if the rocket reaches too great a speed? If it is moving fast enough, the rocket will continue moving away from Earth even after using up its fuel. If a rocket's speed is just right, it will go into orbit around Earth. That happens when the forward momentum just equals the pull of gravity.

When a rocket is in orbit, it is actually falling toward Earth, but just missing it. (The orbit is the path of one object moving around another object.)

Newton explained orbital motion with his law of universal gravitation. He worked out the principles as he tried to understand why the Moon did not fall to Earth like an apple.

He reasoned that all objects in the universe attract one another. The force of gravity, Newton theorized, pulled the Moon toward Earth. Without the pull of Earth's gravity, the Moon would fly off into space. The Moon is continuously falling around Earth, yet flying away from it.

The planets must fall around the Sun in the same way that the Moon falls around Earth. A planet's momentum is equal to the gravitational pull of the Sun on the planet. (The gravity of the rest of the solar system also affects a planet, but that is small compared to the Sun's gravity.)

Take away the Sun and the planets would fly off into space. Newton reasoned that the larger the celestial body, the more gravity it has. He calculated how much the effect of gravity de-

Photo by NASA

creases as bodies move farther apart.

Kepler had determined through observations that planets move in elliptical orbits. Newton explained mathematically why planets move in ellipses. (An ellipse is a closed curve that looks like a flattened circle.)

Newton would surely be pleased by how his ideas are now applied. Today, robot spacecraft use the gravity of one planet to reach another planet. These gravity-assist trajectories enable space probes to increase their speeds. The added speed lets them reach other destinations without using extra rocket fuel.

Newton's theories on gravitation and objects in motion were revolutionary in his time. They have lasted for hundreds of years. The foundation that has made space exploration possible is due to Isaac Newton's innovative mind and sense of wonder.

Social Studies: A Successful Park

Purpose

To determine factors that affect the success of theme parks and amusement parks.

Materials

For each group:
- Atlas of the United States

For each student:
- Drawing compass
- 3 sheets of clear acetate (optional)

Procedure

With your class, list as many factors as you can that might influence the success of a new amusement park. From the class list, choose three factors you think are most important.

The chart below identifies 15 of the top 25 amusement parks in the United States as measured by their 1994 attendance. Some theme parks are themselves destinations. For example, people fly to Orlando, Florida, just to go to Disney World. These kinds of parks have been omitted.

Work as a group to locate the parks on a map of the United States.

Each group member should draw the table shown below on a sheet of paper. Include all columns.

Once you have constructed your table, divide the work of completing it. Try to give each group member the same number of parks to measure. For each park, list all major cities within the distances listed. Use a compass to check the distances on a map of the United States. Be careful not to draw on the map. Once all members have gathered the data for their parks, share answers and complete the table.

Rank	Park	Location	1994 Attendance*
8	Knott's Berry Farm	Buena Park, CA	3.8
11	Cedar Point	Sandusky, OH	3.6
12	Six Flags Magic Mountain	Valencia, CA	3.5
13	Paramount's Kings Island	Kings Island, OH	3.3
14	Six Flags Great Adventure	Jackson, NJ	3.2
16	Six Flags Over Texas	Arlington, TX	3.0
17	Six Flags Great America	Gurnee, IL	2.9
19	Six Flags Over Georgia	Atlanta, GA	2.6
20 (tie)	Paramount's Great America	Santa Clara, CA	2.5
20 (tie)	Knott's Camp Snoopy	Bloomington, MN	2.5
22 (tie)	Six Flags Astroworld	Houston, TX	2.4
22 (tie)	Paramount's King's Dominion	Doswell, VA	2.4
24	Busch Gardens The Old Country	Williamsburg, VA	2.3
25 (tie)	Opryland	Nashville, TN	2.0
25 (tie)	Fiesta Texas	San Antonio, TX	2.0

Source: Amusement Business magazine
* in millions

Conclusion

Where are successful amusement parks located? Write your answer on a sheet of lined notebook paper. Use details from your table to support your answer.

As a group, select another town or city where a new amusement park might be successful. Give three or more sensible reasons for placing your new amusement park there.

Park	Cities Within 50 Miles	Cities Within 100 Miles	Cities Within 200 Miles

Math: Warm-Up Activities

Purpose
To practice math skills using word problems about an amusement park.

Procedure
Find answers for the following problems as directed by your teacher:

1. An amusement park will reduce its admission price 20 percent for students who get As in both science and mathematics. What will the discount price be for a $27.00 ticket?

2. The Mondo Mountain ride takes 6 minutes. The Mellow River cruise takes 16 minutes. If both rides start at 10:00 a.m., when is the next time they will both start at the same time?

3. The acceleration (a) of a mass (m) by a force (F) can be expressed by the formula $a = F/m$. Find the acceleration when the force is 25 newtons and the mass is 2.5 kilograms. (The answer will be in meters per second per second.) What happens to the acceleration if you double both the force and the mass?

4. We can find the distance a ball travels by using the formula $S = vt$, where S represents distance, v is velocity, and t represents time. Find the distance traveled when a ball rolls for 10 seconds at a velocity of 2 feet per second. What happens to the distance when the time is doubled?

5. The Reverse Rotor ride has a radius (r) of 12 feet. How many people can ride if each person standing around the circumference (C) of the ride must be given 3 feet of space? ($C = 2\pi r$)

6. The park must maintain a 99.5 percent or better accident-free ride record in order to renew its license each year. Last year the park reported 828 accidents over a period of 180,000 rides. What percentage of the rides are accident free? Will the park have its license renewed?

7. The waiting area for Tetrahedral Spin is too small. If the length of the waiting area is increased by 25 percent and the width is increased by 20 percent, the area will increase by what percentage?

Math: Turn Off the Rides

Purpose

To use the techniques of simplifying and making a table to solve a problem.

Background

The City Power Company (CPC) has asked businesses to conserve energy on extremely hot days. As manager of an amusement park, you want to follow CPC's recommendations. If you and other managers don't help conserve, CPC may cut off power to all businesses.

Your assistant has a plan to randomly select which of the park's 100 rides you will close. Her plan is to leave all 100 rides turned off for the first minute. Then, turn on every second ride. After another minute, every third switch is changed—if it is off, it is switched on, and if it is on, it is switched off. When another minute passes, every fourth switch is changed and so on until 100 minutes have elapsed. After 100 minutes, each ride that is off would remain off for the rest of the day. On the second day, you can try a different plan.

What a crazy idea! However, since you do not have a better one, you decide to give it a try.

Procedure

There must be a faster way to come up with the same answer. Make the problem simpler by looking for any patterns that appear after ten minutes. Make a table showing the position of the first ten switches each minute for the first ten minutes. Identify which rides are closed after ten minutes. Find what these numbers have in common. Use the pattern you find to identify the rides that will remain closed for the day.

Conclusion

Which rides were closed after 100 minutes? What percentage of the rides were closed? If the park opens at 10:00 a.m., and it takes 5 minutes to place signs on the closed rides, find the latest time you could start and be ready at opening time. How many rides would be closed if the park had 200 rides? What percentage of the 200 rides would be closed? Give at least two reasons why your assistant's method would be less likely to be used if the park had 1000 rides.

English: Thrilling Memories

Purpose

To write a description of your last trip to an amusement park.

Procedure

Remember the last time you visited a carnival, an amusement park, a fair, or similar place. Recall the sights and sounds you experienced there. Think of the things you smelled and some of the special things you tasted.

You are going to write a description of that carnival or amusement park for your school newspaper. The editor wants you to create a vivid word picture for the readers.

Before you begin writing, list the things you remember about the trip. As you complete this prewriting activity, think about how each of your five senses experienced the day.

Imagine you are there right now. What do you see as you look around? If you close your eyes, what sounds and smells stand out in your memory? What do you feel underfoot or against your skin? Finally, what special foods do you remember? How did they taste? If you went on a thrill ride, what thrills did you get? How did the ride make you feel?

When your list is complete, think of how you can link the words in the list together. Create a web or other graphic organizer. As you move from the word list to the web, focus on a small part of the whole story. Do not try to describe every part of your day. Instead, select one ride or experience that captures the feeling of the place.

Now write your first draft. Try to include as many specific, vivid details as possible. Select verbs that best describe what is happening. Organize details in such a way that your reader can easily envision what you are describing.

Conclusion

You might want to give your description to a friend to review. Ask for constructive criticism and revise your work before you write the second draft. Continue to do this until you are satisfied you have the best possible description.

Writing to Persuade

Purpose

To demonstrate knowledge of Newton's Laws of Motion in a persuasive-style letter to the president of Goround Amusements, Inc.

Background

Your company has just designed a ride for the new amusement park, Sir Isaac's Inertialand. Although you are only required to demonstrate a model of your design, a letter to the president of the company will help to strengthen your case.

Materials

- Discovery File "Newton's Laws, A Moving Experience" (page 23)
- Discovery File "Three Laws for the Price of One" (page 40)
- Discovery File "The Force is With You" (page 30)
- Copy of Peer-Response Form (page 53)
- Proofreading Guidesheet (page 54)
- Student Voices

Prompt

As the proud designer of a new thrill ride, you now have the job of persuading the president of Goround Amusements to choose your ride for the new park being built on the outskirts of town. You have decided to write a strong letter emphasizing the features of your ride and stressing the ways it demonstrates Newton's Laws of Motion. You want your letter to persuade the president that your ride will demonstrate each of Newton's Laws, and will be fun and safe to ride.

Begin with an introduction that gets Ms. Goround's attention. Describe your ride and its target audience. Tell about the points in your ride where each of Newton's Laws is demonstrated. Also describe the safety features of your ride. Use facts and data from science activities and discovery files to support your statements.

Finally, write a brief conclusion that sums up your points and tells whom to call if Ms. Goround has questions.

An exceptional letter:
- gets the reader's attention.
- clearly states an opinion.
- gives facts to support the opinion.
- has a strong conclusion.
- has no errors in grammar, punctuation, or spelling.

Use the Proofreading Guidesheeet on page 54 to edit your letter. Have your peers evaluate and react to your letter using a copy of the Peer-Response Form on page 53.

Questions

1. How do you get your reader's attention?
2. How and where is your opinion stated?
3. What information about Newton's Laws do you include? What other information might you include?
4. What data and facts do you use to support your opinion?
5. What other data might you use?
6. How could you improve your conclusion to make it stronger?

Peer-Response Form

Directions

1. Ask your partners to listen carefully as you read your rough draft aloud.

2. Ask your partners to help you improve your writing by telling you their answers to the questions below.

3. Jot down notes about what your partners say:

   a. What did you like best about my rough draft?

   b. What did you have the hardest time understanding in my rough draft?

   c. What can you suggest that I do to improve my rough draft?

4. Exchange rough drafts with a partner. In pencil, place a check mark near any spelling, punctuation, or grammatical construction about which you are uncertain.

5. Return the papers and check your own. Ask your partner about any comments you do not understand or agree with on your paper. Jot down notes you want to remember when writing your revision.

Proofreading Guidesheet

1. Have you identified the assigned purpose of the writing assignment? Have you accomplished this purpose?

2. Have you written on the assigned topic?

3. Have you identified the assigned form your writing should take? Have you written accordingly?

4. Have you addressed the assigned audience in your writing?

5. Have you used sentences of different lengths and types to make your writing effective?

6. Have you chosen language carefully so the reader understands what you mean?

7. Have you done the following to make your writing clear for someone else to read?

    • used appropriate capitalization

    • kept pronouns clear

    • kept verb tense consistent

    • used correct spelling

    • used correct punctuation

    • used complete sentences

    • made all subjects and verbs agree

    • organized your ideas into logical paragraphs

Video

America's Greatest Roller Coaster Thrills, Telemedia Productions, 1994, Goldhil Video, Thousand Oaks, CA

Organizations

American Coaster Enthusiasts (ACE), P.O. Box 8226, Chicago, IL 60680

International Association of Amusement Parks and Attractions (IAAPA), 1448 Duke Street, Alexandria, VA 22314.

Books

Adams, Judith A. *The American Amusement Park Industry—A History of Technology and Thrills.* Boston, MA: Twayne Publishers, 1991.

Cartmell, Robert. *The Incredible Scream Machine—A History of the Roller Coaster.* Bowling Green, OH: Bowling Green State University Popular Press, 1987.

Epstein, Lewis Carroll. *Thinking Physics Is Gedanken Physics—Practical Lessons In Critical Thinking* (Second Edition). San Francisco, CA: Insight Press, 1995.

Freeman, Ira. (Revised William J. Durden.) *Physics Made Simple* (Revised Edition). New York: Bantam Doubleday Dell Publishing Group, Inc., 1990.

Kasson, John F. *Amusing the Million—Coney Island at the Turn of the Century.* New York: Hill and Wang, a Division of Farrar, Straus and Giroux, 1978.

Kuhn, Karl F. *Basic Physics—A Self-Teaching Guide.* New York: John Wiley & Sons, Inc., 1979.

Mungenast, Marcia. *Kings Dominion Science Day Middle/Secondary Teacher Resource.* Doswell, VA: Paramount's Kings Dominion, Published by Prentice-Hall, Inc., 1993.

O'Brien, Tim. *The Amusement Park Guide.* Chester, CT: The Globe Pequot Press, 1991.

Paramount's Great America Physics Day—Science Day 1995. Santa Clara, CA: Paramount's Great America, 1995.

Throgmorton, Todd. *Roller Coasters of America.* Osceola, WI: Motorbooks International Publishers & Wholesalers, 1994.

Wiese, Jim. *Roller Coaster Science—50 Wet, Wacky, Wild, Dizzy Experiments About Things Kids Like Best.* New York: John Wiley & Sons, Inc., 1994.

Kits

Amusement Park Physics Accelerometer Kits
PASCO Scientific
10101 Foothills Blvd.
Roseville, CA 95678
Phone: 1-800-772-8700
(Call for kit prices and ordering information)

ACKNOWLEDGMENTS

Author

Russell G. Wright, with contributions from Leonard David, Barbara Sprungman, Janet Wert Crampton, and the following teachers:

Science Activities

*William R. Krayer, Gaithersburg High School, Gaithersburg, MD
*Frank S. Weisel, Tilden Middle School, North Bethesda, MD

Teacher Advisors

Patricia L. Berard, Colonel E. Brooke Lee Middle School, Silver Spring, MD
Henry Milne, Cabin John Middle School, Potomac, MD
Eugene M. Molesky, Ridgeview Middle School, Gaithersburg, MD
Kenneth A. Schmidt, Redland Middle School, Rockville, MD
Barbara L. Teichman, Parkland Middle School, Rockville, MD

Interdisciplinary Activities

*James J. Deligianis, Tilden Middle School, North Bethesda, MD
*Cuyler J. Cornell, Cabin John Middle School, Potomac, MD

Event/Site Support

Jeff Duncan, Culpeper County Middle School, Culpeper, VA

Scientific Reviewers

Cynthia Emerick, Togo International, Inc.
Ron Toomer, Arrow Dynamics
Walt Davis, Togo International, Inc.
Robert Speers, Firelands College

Student Consultants

*Redland Middle School, Rockville, MD:
Norah Harwood, Peter Ebert, Shani Ramadhan, Jennifer Wallen, Michael Preissner, Ian Giles, Jennifer Liang, Carlo Cordero, Thy Nguyen, Pamela Finder, Dehkonti Paelay, Melissa Carter
*Tilden Middle School, Rockville, MD:
Jeffrey Hall, William Wright

Field-Test Teachers

Patricia Flynn, Agawam Jr. High, Feeding Hills, MA
Cheryl Glotfelty and Linda Mosser, Northern Middle School, Accident, MD
Gerry Harrison, Sevier Middle School, Kingsport, TN
Hope Hall, Ross N. Robinson Middle School, Kingsport, TN
Lilla Green, Hartigan School, Chicago, IL
Judy Kidd, South Charlotte Middle School, Charlotte, NC
Helen Linn and Cathy Miller, Huron Middle School, Northglenn, CO
Wendy Beavis and Mike Geil, Tanana Middle School, Fairbanks, AK

EBS Advisory Committee

Dr. Joseph Antensen, Baltimore City Public Schools
Ms. Deanna Banks-Beane, Association of Science-Technology Centers
Ms. MaryAnn Brearton, American Association for the Advancement of Science
Dr. Jack Cairns, Delaware Dept. of Public Instruction
Mr. Bob Dubill, *USA Today*
Mr. Gary Heath, Maryland State Department of Education
Dr. Henry Heikkinen, University of Northern Colorado
Dr. J. David Lockard, University of Maryland (Emeritus)
Dr. Ramon Lopez, American Physical Society
Dr. Wayne A. Moyer, Montgomery County Public Schools (Retired)
Dr. Arthur Popper, University of Maryland

*Asterisks indicate Montgomery County Schools